Schriftenreihe Keramische Werkstoffe
Lehrstuhl Keramische Werkstoffe

Herausgeber Prof. Dr.- Ing. Walter Krenkel

Band 12

I0042184

# Thermal barrier coating by polymer-derived ceramic technique for application in exhaust systems

Von der Fakultät für Ingenieurwissenschaften

der Universität Bayreuth

zur Erlangung der Würde eines

Doktor-Ingenieurs (Dr.-Ing.)

vorgelegte Dissertation

vorgelegt von

Eng. Gilvan Barroso

aus

Blumenau, Brasilien

Erstgutachter: Prof. Dr.-Ing. Walter Krenkel

Zweitgutachter: Prof. Dr. rer. nat. Michael Scheffler

Tag der mündlichen Prüfung: 21. November 2017

Lehrstuhl Keramische Werkstoffe

Universität Bayreuth

2017

**Bibliografische Information der Deutschen Nationalbibliothek**

Die Deutsche Nationalbibliothek verzeichnet diese Publikation in der Deutschen Nationalbibliografie; detaillierte bibliographische Daten sind im Internet über http://dnb.d-nb.de abrufbar.

1. Aufl. - Göttingen: Cuvillier, 2018
    Zugl.: Bayreuth, Univ., Diss., 2017

© CUVILLIER VERLAG, Göttingen 2018
    Nonnenstieg 8, 37075 Göttingen
    Telefon: 0551-54724-0
    Telefax: 0551-54724-21
    www.cuvillier.de

1. Auflage, 2018
Gedruckt auf umweltfreundlichem, säurefreiem Papier aus nachhaltiger Forstwirtschaft.

    ISBN 978-3-7369-9832-2
    eISBN 978-3-7369-8832-3

# Table of Contents

1 Introduction, Motivation and Objectives ....................................................... 4

2 Background and Literature Overview ............................................................ 7

  2.1 General aspects of polymer-derived ceramics (PDCs) .......................... 7

    2.1.1 Preceramic polymers ............................................................................ 8

    2.1.2 Polymer-to-ceramic conversion ....................................................... 10

    2.1.3 PDC processing with fillers .............................................................. 13

  2.2 Polymer-derived ceramic coatings ........................................................ 16

    2.2.1 Processing of PDC coatings .............................................................. 17

    2.2.2 State of the art of particle-filled PDC coatings ............................. 25

  2.3 Thermal barrier coatings ......................................................................... 33

    2.3.1 Materials of interest for the top-coat .............................................. 35

    2.3.2 Thermal properties and failure mechanisms ................................ 37

  2.4 Automotive exhaust systems ................................................................... 40

    2.4.1 Automotive emissions and European legislations ........................ 41

    2.4.2 Methods for emissions control ......................................................... 44

3 Experimental Procedures ............................................................................. 48

  3.1 Selection of materials ............................................................................... 48

    3.1.1 Substrate .............................................................................................. 48

    3.1.2 Preceramic polymers ......................................................................... 50

    3.1.3 Fillers ................................................................................................... 52

    3.1.4 Solvent and Dispersant ..................................................................... 53

  3.2 Compositions and preparation of coating suspensions ...................... 54

  3.3 Deposition of the coatings ....................................................................... 56

  3.4 Thermal treatment .................................................................................... 58

  3.5 Characterizations ...................................................................................... 60

    3.5.1 Behavior during thermal treatments .............................................. 60

    3.5.2 Thermal expansion ............................................................................ 61

    3.5.3 Thermal conductivity ........................................................................ 62

    3.5.4 Thermal insulation ............................................................................ 65

3.5.5 Surface roughness.................................................................................................66

3.5.6 Coating thickness................................................................................................67

3.5.7 Imaging techniques ............................................................................................67

3.5.8 Density.................................................................................................................68

3.5.9 Composition ........................................................................................................68

3.5.10 Adhesion............................................................................................................69

3.5.11 Stress evolution.................................................................................................71

3.5.12 Temperature and oxidation resistance..........................................................72

3.5.13 Thermal shock resistance ...............................................................................73

4 **RESULTS AND DISCUSSIONS** ...........................................................................................74

**4.1 Characterization of the selected materials**............................................................74

4.1.1 Properties of the substrate................................................................................74

4.1.2 Conversion behavior of the coating components ...........................................76

**4.2 Development of the coating system**.........................................................................80

4.2.1 Bond-coat ............................................................................................................80

4.2.2 Top-coat...............................................................................................................83

**4.3 Conversion behavior of the TBC system** ................................................................87

4.3.1 Mass change........................................................................................................87

4.3.2 Dimensional changes .........................................................................................88

**4.4 Microstructure**...........................................................................................................96

4.4.1 Morphology of the cross-section .....................................................................96

4.4.2 Morphology of the surface ..............................................................................102

**4.5 Coating adhesion**....................................................................................................102

**4.6 Thermal properties**.................................................................................................106

4.6.1 Coefficient of thermal expansion...................................................................106

4.6.2 Thermal conductivity ......................................................................................108

**4.7 Durability**................................................................................................................109

4.7.1 Thermal shock resistance................................................................................109

4.7.2 Long-term temperature and oxidation resistance........................................110

**4.8 Coating on the inside of pipes** ..............................................................................116

4.8.1 Deposition of coatings.....................................................................................116

4.8.2 Investigation of the insulating effect.............................................................117

5   SUMMARY ...... 119

6   OUTLOOK ...... 123

7   ZUSAMMENFASSUNG ...... 125

8   REFERENCES ...... 130

9   ACKNOWLEDGEMENTS ...... 149

10  ANNEXES ...... 150

    10.1  List of Symbols, Variables, Chemical Compounds and Abbreviations ...... 150

    10.2  Publications ...... 154

    10.3  Curriculum Vitae ...... 155

# 1 Introduction, Motivation and Objectives

A great step in the technological development of civilizations was the combination of different materials to obtain the desired final properties. In this context, the development of coatings stands out, as they can modify substrates to change appearance, functionality and/or increase resistance against the environmental conditions. Due to their intrinsic properties, polymeric, metallic and ceramic coatings are suitable for different applications. However, not only the properties, but also the processing plays an important role in the choice of the coating materials. As a consequence, even though a given material may offer the best properties for the intended application, the processing of this material as a coating may be too difficult and expensive, or even impossible. Indeed, in more extreme conditions, like harsh environments and high temperature applications, the hall of materials with sufficient resistance is limited, and ceramic coatings are often the only suitable option. Contradictorily, the same properties – like high hardness and stability up to very high temperatures – make the processing of these coatings a great technological challenge, requiring generally expensive and/or time-consuming processes.

Another characteristic of the processing of ceramic coatings, which increase the difficulty of the processing, are the synergistic effects resulting from interactions between coating, substrate and environment. These effects become even more relevant when coatings are deposited onto non-ceramic substrates. Good examples of such systems are thermal barrier coatings (TBCs) [PaGJ2002]. These coatings provide the surface of metallic substrates – in general, with lower temperature resistance – properties of refractory ceramics. Due to the low thermal conductivity of these coatings, the temperature of the substrate is reduced, enabling the expansion of application limits of the metal and extension of the lifespan of metallic parts.

The predecessors of TBCs were enamel coatings developed for military engines in the 1950s. However, it was only two decades later that TBCs were successfully applied in gas turbine to protect parts against damages caused by the high temperatures [Mill1997]. This was an important technological improvement for aerospace and energy industries, as an enhanced temperature resistance of the parts enabled operation of turbines with higher inlet temperatures, resulting in a significant increase of turbine efficiency. Since then, application of TBCs has expanded significantly [KuKa2016]. In military and aerospace sectors, TBCs have also been used on wings and nose of rockets and missiles, in order to protect these parts against heat generated by dislocation at extreme velocities, and at the inside of combustion chambers. The automotive industry has been applying TBCs for two main reasons. The first relates to the performance of engines, where TBCs protect the combustion chambers and piston tops, enabling higher combustion temperatures without damages to the engine. Higher combustion temperatures are associated with higher efficiency and reduction of emissions. The second application relates to the protection of parts surrounding hot components. In this case, TBCs deposited on the exterior side of hot components reduce heat transfer to the environment, protecting less temperature-resistant parts in the surroundings from excessive heat.

In aerospace, energy and military sectors, priority is to obtain effective, reliable and durable coatings, whereas processing cost is a secondary aspect. For automotive applications, however, processing cost is decisive and limits the range of application of TBCs. Hence, although very effective, the use of TBCs by the automotive industry is still limited to isolated branches, like racing and military vehicles, and to high-end segments.

A third possible automotive application aims the increase of efficiency of exhaust systems by depositing TBCs on the inside of exhaust pipes. The advent of internal combustion engines brought with it great concerns about air pollution, especially from the 1950s, when photochemical smog and acid rain have become frequent occurrences in large cities. Several technologies have been implemented – e.g. three-way catalytic converters – and automotive emissions reduced drastically since then. Nevertheless, legislations worldwide have been frequently reducing emission limits further, forcing the automotive industry to continuously search for ways to reduce these emissions. Catalytic converters are currently able to convert sufficiently gaseous pollutants during normal operation, in order to fulfill legislation requirements. However, the conversion rate in a catalytic converter is dependent on temperature. Hence, when the system is cold, catalytic converters are not able to convert gaseous pollutants sufficiently. Indeed, the so-called cold start behavior is responsible for the majority of automotive emissions [Reif2015]. The low thermal conductivity of a thermal barrier coating applied on the inside of pipes preceding the catalytic converter would reduce heat loss from the exhaust gas to the metallic pipes during the first seconds of operation. Thus, combustion gases would reach the catalytic converter with higher temperature and sufficient conversion rates would be achieved faster.

Together with high processing cost, feasibility of coating deposition onto the inside of pipes is a limiting factor, especially in the case of long pipes with small inner diameter. The APS method requires a plasma gun with much bigger dimensions than typical exhaust pipe diameters. EB-PVD, on the other hand, is a deposition technique in gas phase. Hence, only exposed surfaces can be coated homogeneously, which is not the case of the inside of pipes.

In this context, the development of a thermal barrier coating by polymer-derived ceramic (PDC) technique is proposed. The PDC route enables deposition of silicon-based polymeric coatings in liquid phase by simple lacquer techniques. After deposition, a thermal treatment induces cross-linking and transformation of preceramic polymers into ceramic materials [CMRS2010]. This strategy enables a simple and relatively cost-effective processing, and offers improved flexibility regarding substrate's geometry, which could open doors not only of the automotive industry, but also from other segments, to the use of TBCs in large scale.

To obtain a coating system with high thickness, strong adhesion, low thermal conductivity and high thermomechanical stability is the greatest challenge of this development, since most PDC-based coatings on metallic substrates are limited to a few microns in thickness and/or to application temperatures up to ~800 °C. The coatings developed in this work should have a thermal conductivity comparable to conventional TBCs, sufficient thermal stability for application in exhaust systems, and a large thickness obtained by a single deposition step to reduce processing time. The thermal conductivity must remain low up to temperatures sufficient for the catalytic converter to achieve full conversion of the pollutants, i.e. about 500 °C [YuKi2013, Reif2015]. Furthermore, to enable application in exhaust systems, deposition on the inside of exhaust pipes must be feasible.

In automotive exhaust systems, maximum temperatures of ~950 °C may occur. Thus, the developed coatings must be able to withstand such conditions. However, exhaust systems of conventional vehicles reach temperatures above 900 °C only sporadically and for a few seconds at a time. In fact, exhaust system developers consider average temperatures of 700-750 °C as typical for exhaust pipes located between manifold and catalytic converter. Thus, although the coating system must be able to withstand temperatures as high as 950 °C for short periods, they must resist long-term exposure to temperatures of at least 750 °C. Moreover, severe temperature variations occur in exhaust systems, especially during winter or rainy days, due to water splashes. Hence, the coatings must also withstand thermal shock of at least 700-800 °C, corresponding to the immediate cooling from normal operation conditions to negative ambient temperatures.

## 2 BACKGROUND AND LITERATURE OVERVIEW

This chapter presents a background for the development of thermal barrier coatings by PDC-processing with focus on application in automotive exhaust systems. The following sections offer an overview about PDC technology with emphasis on coatings, on conventional processing and properties of thermal barrier coatings, as well as on exhaust systems and automotive emissions.

### 2.1 General aspects of polymer-derived ceramics (PDCs)

The polymer-derived ceramic route is an alternative method for preparation of ceramic materials. It is based on the conversion of suitable molecular precursors, generally called preceramic polymers, into ceramics by a series of thermally and/or chemically-induced processes. The resulting ceramics may be binary materials like $Si_3N_4$, SiC, BN, and AlN; ternary such as SiCN, SiCO, and BCN; or even quaternary like SiCNO, SiBCN, SiBCO, SiAlCN, and SiAlCO [CMRS2010]. In fact, some of these phases – e.g. SiCN – can only be synthesized homogeneously by PDC processing [PeVB1990].

In comparison to traditional methods for preparation of ceramics, like powder sintering, the PDC-route offers advantages such as lower processing temperature and time, better adjustment of properties, homogeneity in a molecular level, and chemical purity. Notwithstanding, the most interesting characteristic of this processing route is the large variety of shaping techniques, achieving in several cases near-net-shaped parts before any treatment at high temperatures. Preceramic polymers may be formed into complex and/or highly porous 3D parts [CoHe2002, Motz2006], coatings [ToBo2008b, GKDD2009] and fibers [MBCT2007, FBNK2014] through the most varied methods, including typical techniques for polymer processing.

The work of Fritz and Raabe in the 1950s [FrRa1956], reporting about conversion of polycarbosilanes into SiC-based ceramics, is considered the starting-point of the PDC technology. In the following decade, several other studies on the preparation of ceramic materials from molecular precursors were published, among which those of Ainger and Herbert [AiHe1960] and Chantrell and Popper [ChPo1964] standout. In the beginning of the 1970s, first approaches combining the use of molecular precursors and shaping techniques were developed in Germany [Verb1973, VeWi1974, WiVM1974]. Short after, polymer-derived ceramic fibers were developed in Japan [YaHO1975, YHOO1976]. Since then, the PDC technique evolved and the number of applications increased tremendously.

Due to several advantages, interest for PDC technology grows from year to year. Hence, the PDC technique has been subject of a large number of review papers and books throughout the last decades [WyRi1984, PeVB1990, BiAl1995, Grei2000, RMHK2006, CMRS2010, CRSK2010, Bern2012, BFPS2014]. Today, progress of the technology relies mainly on synthesis of tailored preceramic polymers, on understanding and modifying parameters for polymer-to-ceramic conversion and on the use of fillers, to enable preparation of complex shapes and to improve and/or modify properties of ceramic products. Thus, these three main topics are discussed with further details in the following sections.

### 2.1.1 *Preceramic polymers*

The characteristics of preceramic polymers have strong influence on the properties of resulting ceramics. Therefore, an extensive variety of ceramic materials can be prepared by PDC processing only by changing or tailoring precursors. Indeed, the modification and synthesis of new preceramic polymers is a recurrent topic of research in the last decades [Seyf1995, RKDR1996, AbGu2004, ZSMK2012, FSKH2013].

To the group of preceramic polymers belong metalorganic compounds, mostly based on silicon with additional elements (e.g. B, N, C, O) in the main chain, with side groups R = hydrogen, alkyl, vinyl, aryl, etc. [Grei2000, RMHK2006], as exemplified in Fig. 2.1.1. Composition of backbone as well as of side groups attached to it will dictate which ceramics may be obtained. They will determine, especially, phase composition and distribution of elements in the ceramic products, as well as the properties of the material as a polymer and as a ceramic. Indeed, properties like chemical and thermal stability, solubility, rheological behavior and optical characteristics may be tailored simply by changing side groups. Side groups also allow adjustment of physicochemical behavior related to cross-linking reactions and to polymer-to-ceramic conversion [BiAl1995, RMHK2006, CMRS2010].

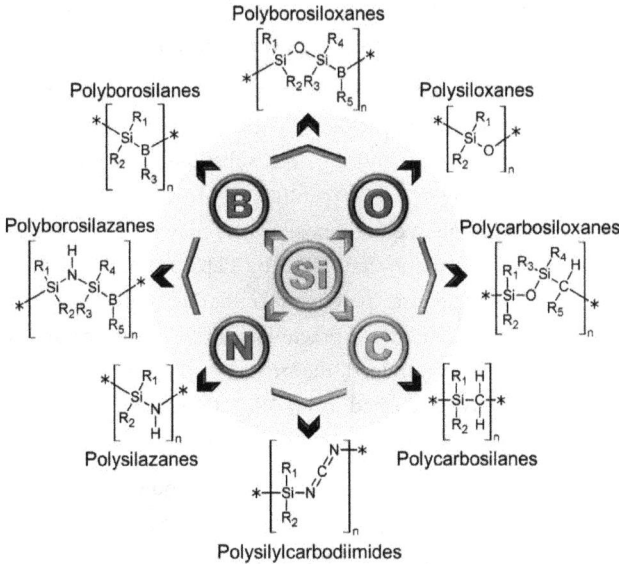

Fig. 2.1.1: Examples of basic units of usual silicon-based preceramic polymers.

Industrial production of silicon-based polymers only became possible in the 1940s after development of the Müller-Rochow process by E. G. Rochow (USA) and R. Müller (Germany) to produce organochlorosilanes. Today, a great variety of these compounds is produced from chlorosilanes, including polycarbosilanes, polysilazanes and polysiloxanes. However, in order to be suitable as precursor for PDC processing, polymers must fulfill some requirements. The first requirement is a sufficiently high molecular weight, to prevent volatilization of light compounds during thermal treatment, enhancing the ceramic yield $\alpha_{pc}$ (defined as the mass residue in percent after polymer-to-ceramic transition related to the initial mass of precursor) [CMRS2010]. During the last years, a great amount of novel polymers were developed, which can provide ceramic yields of 90% or more [Grei2000]. The second requirement is a latent reactivity. The presence of specific chemical groups, which upon curing and cross-linking form branched structures, ensure such reactivity, turning fusible or liquid polymers into thermoset materials. This latent reactivity is normally related to unsaturated organic groups (e.g. vinyl) or to highly reactive bonds such as Si-H or N-H [WyRi1984]. The third requirement is an appropriate and controllable rheology and/or solubility to allow shaping of the polymer for the respective application (bulk parts, fibers, coatings, porous materials, etc.). Precursors are liquid or solid, depending on their structure and molecular weight. If they are solid, they must be soluble or meltable at low temperatures, usually below 150 °C, in order to be suitable for PDC processing. Processing in liquid phase enables application of polymer shaping methods to obtain near-net-shaped parts, reducing finishing work in comparison to powder technology [Grei1998].

The most famous materials among preceramic polymers are the silicones, which belong in the class of polyorganosiloxanes. These polymers – backbone composed of Si and O atoms, with hydrogen and organic substituents – have been studied and synthesized since beginning of the 20th century [KiLl1901]. Additionally to outstanding properties in polymeric state, polyorganosiloxanes may also be used as precursors for preparation of ceramics in the system SiCO by PDC processing [BlMK2005]. Another well-known class of preceramic polymers is that of polycarbosilanes, with -Si-C-backbone. Polycarbosilanes are usual precursors for SiC-based ceramics by means of a pyrolytic process [RRBB1997]. Indeed, the precursor used by Yajima and coworkers in their breakthrough study published from 1975 [YaHO1975] about development of polymer-derived ceramic SiC fibers was a polycarbosilane.

Polymers with Si and N in the backbone – called polysilazanes – or with Si, C and N, like polycarbosilazanes and polysilylcarbodiimides are interesting precursors for PDC processing, especially to prepare SiCN-based materials [SeWi1984, DrRi1997, RGMD1997, GRSM1997, RKGG1998, ZKMM1999, Motz2006, Luka2007]. In fact, the pioneering work of Verbeek and co-workers in the 1970s proposed the use of polycarbosilazanes to obtain small diameter SiCN ceramic fibers [Verb1973, VeWi1974, WiVM1974].

Other molecular precursors have been studied and used for preparation of ceramic materials by PDC processing. Among them, polysilazanes, polysilylcarbodiimides and polysilanes modified with boron or aluminum standout and were used for preparation of SiBCN [RKDR1996, BKMW2001], SiAlCN [BWAM2004, MUKS2006] and SiAlON [SRCB1991, PBCL2014] ceramics. Furthermore, preparation of ceramics from preceramic polymers modified with transition metals has been carried out as well. Especially tailoring with late transition metals (e.g. Fe, Co, Ni, Pd, Pt, Cu, Ag and Au) is of utmost relevance due to enhanced magnetic properties and potential applications in catalytic systems [YIYO1981, BaSo1991, GSKH2010, ZaMK2011, ZSMK2012].

### 2.1.2 Polymer-to-ceramic conversion

Polymer-to-ceramic conversion is a complex process, which involves a series of transformations responsible for drastic structural changes in precursors. As mentioned earlier, one of the fundamental requirements of preceramic polymers is to have a latent reactivity. Products of precursor syntheses – reactions of halogenosilanes by hydrolysis, ammonolysis, etc. – are usually mixtures of oligomers and low molecular weight polymers, which may easily volatilize and depolymerize, resulting in low ceramic yields.

To avoid volatilization, precursors must undergo cross-linking reactions, which increase the molecular weight prior to ceramization [SeWi1984, BlSL1989]. Cross-linking occurs through thermally- or catalytic-induced condensation and/or addition reactions of certain functional groups, such as Si-H, Si-OH, Si-vinyl [IcTI1987, CMRS2010, MSTK2012]. Although in some extent necessary, an excessive cross-linking might prevent further shaping [WWPK2004]. Therefore, control of cross-linking reactions must be taken carefully into consideration during PDC processing. When complete, cross-linking of a suitable precursor results in a solid thermoset polymer, which will not decompose or deform significantly during ceramization [LaBa1993, BiAl1995]. This enables machining of parts before ceramization, reducing wear of tools and avoiding damage of products during finishing process [RGBB2005, Motz2006]. Aside from thermic/catalytic methods, cross-linking may be carried out by several other processes, including UV [SBGH2004, PKKS2007], laser [FTNS2005], electron or γ-ray radiation curing [INOS2001, ISTN2004, FSKH2013], plasma [Lipo1988] and other reactive environments [RaLL1990, Hase1992, PGDM2008].

The next transition is called ceramization and involves thermolysis (generally called pyrolysis) and evaporation of organic groups, which cause the organic-to-inorganic transformation of precursors, resulting in amorphous ceramics [SoBM1988]. Rearrangement and radical reactions initiate at temperatures above 300 °C, resulting in cleavage of chemical bonds and release of organic functional groups ($CH_4$, $C_6H_6$, $CH_3NH_2$, $NH_3$, etc.) [Grei2000, MSTK2012]. Also pyrolysis may be carried out by different methods, like hot pressing [IGBA2002], spark plasma sintering (SPS) [SSKM2015], chemical vapor deposition (CVD) [BSMH2001], plasma spraying [KrUl2006], radiation pyrolysis [ChMa1991], laser [MHTA2003], microwave [DSCP2000], ion irradiation [PiCS2000], and others. Some of these methods even combine shaping and pyrolysis processes in one single step. Nevertheless, pyrolysis in furnace using a suitable atmosphere is still the most common method.

The polymer-to-ceramic conversion is usually completed below 1100 °C [CMRS2010]. However, ideal conditions for cross-linking and ceramization processes depend strongly on the chemical structure of the precursor. Furthermore, by changing conditions for cross-linking and/or ceramization (temperature, heating rate, atmosphere, etc.), different materials may be obtained [BPGC1993].

Amorphous PDCs may be further treated to originate crystalline ceramics. For a great number of applications, crystallization is not necessary, frequently even avoided, and processing is terminated in amorphous stage. Indeed, some PDCs have the capability of avoiding crystallization at temperatures up to 1700 °C [RKDR1996], whereas others – especially ceramics containing oxygen – may crystallize at temperatures as low as 1000-

1200 °C [Grei2000]. One example of a desired crystallization is the preparation of polycrystalline $Si_3N_4/SiC$ nanocomposites from preceramic polymers, which show better properties than mixtures of the individual ceramics [WKSM1990, Niih1991, RKSA1995, RGMD1997]. Fig. 2.1.2 summarizes the general correlation between PDC processing stages [CMRS2010] and precursor's mass change with the temperature.

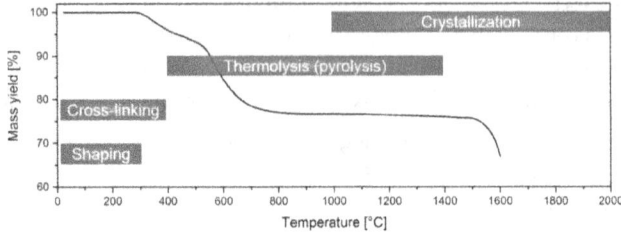

Fig. 2.1.2: *Temperature range of the transformations during PDC processing and example of mass change with temperature (polycarbosilazane under inert atmosphere).*

The PDC route enables preparation of different ceramics at lower temperatures compared to powder sintering process. However, the polymer-to-ceramic transition is associated to major changes in the materials. To evaluate these changes, two important parameters are used as references:

$$\alpha_{pc} = \frac{m_c}{m_p} \qquad \text{(Eq. 2.1.1)}$$

where $\alpha_{pc}$ is the ceramic yield ($m_c$ and $m_p$ are the masses of ceramic and polymer, respectively), and

$$\beta_{pc} = \frac{\rho_p}{\rho_c} \qquad \text{(Eq. 2.1.2)}$$

where $\beta_{pc}$ is called density ratio ($\rho_c$ and $\rho_p$ are the densities of ceramic and polymer, respectively). A mass loss of typically 10-30% [Grei2000] occurs during polymer-to-ceramic transformation ($\alpha_{pc} < 1$). Furthermore, a drastic densification of the material (from about 1 g cm$^{-3}$ as precursor to 2-3.5 g cm$^{-3}$ as ceramic) takes place ($\beta_{pc} < 1$). Hence, $\alpha_{pc}\beta_{pc} < 1$ for all PDC systems known. This means that the volume of material reduces during the polymer-to-ceramic transition and this volume change results either in formation of porosity or in shrinkage of the shaped part, most commonly both. In early stages of ceramization, a series of open-pore channels is formed, which might close again during thermal treatment – in this case, called transient porosity [Grei2000] – or might be partially retained in the final product as residual porosity. Shrinkage is the greatest drawback of the PDC technology. Assuming isotropic behavior, let the linear shrinkage during polymer-to-ceramic transition $\varepsilon_{pc}^l$ be considered as a function of $\alpha_{pc}\beta_{pc}$ and of the volume fraction of voids $V_v$ as follows:

$$\varepsilon_{pc}^l = 1 - \left(\frac{\alpha_{pc}\beta_{pc}}{1 - V_v}\right)^{1/3} \qquad\qquad \text{(Eq. 2.1.3)}$$

By plotting the linear shrinkage as a function of $\alpha_{PC}$ and $\beta_{PC}$, and $\alpha_{PC}\beta_{PC}$ (Fig. 2.1.3), it becomes clear that a linear shrinkage comparable with powder sintering of a fully dense ceramic (~20%) can only be obtained if $\alpha_{PC}\beta_{PC} > 0.5$, which is rarely the case for PDC systems [GrSe1992].

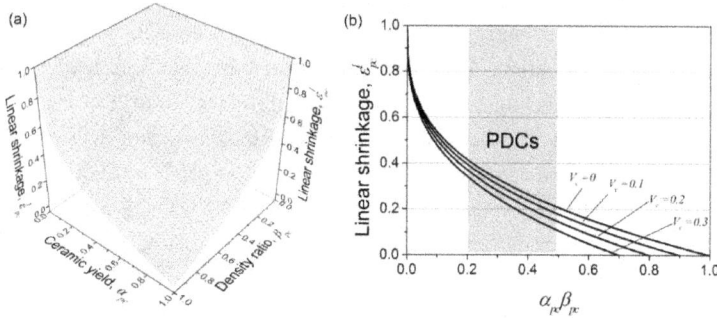

Fig. 2.1.3: Variation of linear shrinkage during polymer-to-ceramic transition ($\varepsilon_{pc}^l$) with the ceramic yield ($\alpha_{pc}$) and the density ratio ($\beta_{pc}$): (a) for a fully dense PDC ($V_v = 0$), and (b) $\varepsilon_{pc}^l$ vs. $\alpha_{pc}\beta_{pc}$ for different volume fractions of voids ($V_v$) in the PDC.

If the material is not able to sufficiently dissipate stresses by viscous flow or diffusional material transport, excessive shrinkage will lead to cracking, which reduces mechanical stability and may cause collapse of the part [Grei1995, RMHK2006]. For these reasons, the preparation of ceramic materials by PDC processing using only preceramic polymers is mostly suitable for shapes with small dimensions like fibers, thin coatings, and cellular ceramics (thin cell wall), which can relieve stresses smoothly [CMRS2010]. Notwithstanding, shrinkage might be reduced or compensated by using suitable fillers, enabling preparation of dense bulk parts and thicker coatings as well.

### 2.1.3  PDC processing with fillers

The use of fillers in PDC processing is a way to overcome shrinkage of precursors during the polymer-to-ceramic transition, enabling near-net-shape manufacturing of dense ceramics virtually without geometric limitations. This strategy offers, in addition, the possibility of modifying properties of the final material, either by introducing new phases or by modifying the microstructure of the PDC.

Fillers may be polymers, ceramics, glasses or metals, and are usually added in form of particles (nano-, submicron-, or micrometric), but also as flakes, nanotubes and fibers. They may be subdivided in four basic types, according to their behavior in the system: passive, active, meltable or sacrificial fillers.

Passive fillers are those materials, which are inert in the system and thus, react neither with precursors nor with the pyrolysis atmosphere (including gaseous products of pyrolysis reactions), retaining the initial mass and particle size (disregarding thermal expansion). These fillers only reduce the overall shrinkage by reducing the volume fraction of the shrinking component, i.e. the preceramic polymer. Passive fillers are mostly ceramics, like SiC, $B_4C$, $Si_3N_4$, $ZrO_2$, BN, $Al_2O_3$, $SiO_2$ and others [ToBo2010]. It is important to mention that some materials might behave as passive fillers only under certain conditions, such as non-oxidative atmosphere and/or lower temperatures, whereas they might behave as an active, a meltable or a sacrificial filler under other conditions. Passive fillers may have the same composition of the PDC, e.g. SiC particles in a polymer-derived SiC matrix [ZHDY2008]. In this case, they will not change the composition of the material but will enable PDC processing by reducing the overall shrinkage. Due to their inert nature, high volume fractions of passive fillers can be added to preceramic polymers, in order to tailor functionality of the final ceramic. Indeed, precursors might be used simply as binders or sintering additives [ScRo1986, ScBG1994] for particles, with the advantages of enhanced form-stability after cross-linking and reduced mass loss, even at high temperatures. Although this strategy may avoid completely the macroscopic shrinkage of the part caused by precursor densification, residual porosity will always occur as a result of the addition of medium or high volume fractions of passive filler, with the amount of residual porosity increasing with increasing volume fraction of filler – assuming filler powder constituted of spherical particles of same size [GrSe1992].

Active fillers, in contrast to the passive counterpart, are not inert in the system. They react either with precursor or with atmosphere (frequently with gaseous products of pyrolysis reactions) to generate new materials. Metallic and semimetallic materials, especially Si, B, Al, transition metals from groups 4 to 6 of the periodic table and their respective silicides and borides, are the most common active fillers [Grei1995, ToBo2010]. In the great majority of cases, such fillers are used to compensate shrinkage of preceramic polymers by formation of phases with a higher molar volume than the initial filler material. Greil and Seibold presented a methodology to calculate the necessary amount of active filler to obtain PDCs with zero shrinkage during ceramization process, without introducing porosity: the active-filler-controlled pyrolysis (AFCOP) method [GrSe1991, GrSe1992]. The governing equation (Eq. 2.1.4) relates the linear shrinkage of precursor during polymer-to-ceramic transition $\varepsilon_{pc}^l$ (Eq. 2.1.3) with the linear expansion of active filler particles $\varepsilon_{af}^l$ (Eq. 2.1.5) – determined by $\alpha_{af}$ and $\beta_{af}$, parameters for mass and density changes, which the active fillers undergo during thermal treatment – to calculate linear shrinkage of particle-filled precursors $\varepsilon_{paf}^l$. Thus, the necessary amount of active filler to obtain a dense ceramic body without shrinkage may be theoretically estimated [GrSe1991, GrSe1992]:

$$\frac{\varepsilon_{paf}^{l}}{\varepsilon_{pc}^{l}} = 1 - \frac{V_{af}}{V_{af}^{*}}\left(1 - \frac{V_{af}^{*}}{\varepsilon_{pc}^{l}}\varepsilon_{af}^{l}\right) \qquad \text{(Eq. 2.1.4)}$$

$$\varepsilon_{af}^{l} = 1 - \left(\alpha_{af}\beta_{af}\right)^{1/3} \qquad \text{(Eq. 2.1.5)}$$

$$V_{af}^{*} = V_{af}^{max}\left(3 - \alpha_{pc}\beta_{pc} - \alpha_{af}\beta_{af}\right) - \left(1 - \alpha_{pc}\beta_{pc}\right) \qquad \text{(Eq. 2.1.6)}$$

where $V_{af}^{*}$ is the critical active filler volume fraction in the starting mixture, which determines the maximum particle packing of active filler after ceramization of the precursor, and $V_{af}^{max}$ is the maximum packing density of filler particles alone and without thermal treatment (0.74 for spherical particles of same size [GrSe1992]). Similar to passive fillers, active fillers may be used to introduce new phases into a PDC matrix, in order to modify functionality and/or performance, together with shrinkage control and densification. In fact, they may even be used to obtain completely new materials by solid-state reactions with precursors, with advantage of low shrinkage during thermal treatment [HHRW1999, STKM2014].

The third type of fillers in PDC systems are the meltable ones, which are mostly glasses. These fillers will change their physical characteristics, by melting or simply softening, during thermal treatment. They may or may not react with other components in the system [OKWB2011]. Therefore, meltable fillers are useful to seal porosity, to enhance hardness of amorphous ceramics and, especially in PDC coatings, to improve oxidation and corrosion protection [GSGW2011]. By softening, they also enable a relaxation of thermomechanical stresses caused by high temperatures and increase the mobility of components in the system, especially when used in combination with passive or active fillers. Despite these benefits, the use of glasses as fillers requires a fine-tuning of the thermal treatment. Too high temperatures may lead to an excessive reduction of the glass melt viscosity, leading to an inhomogeneous distribution of glass within the PDC, and even to crystallization, causing additional shrinkage. An insufficiently high temperature, on the other hand, will not lead to softening/melting of the glass and particles will behave as passive fillers.

Sacrificial fillers are mostly organic compounds, which can be eliminated by thermal decomposition or by dissolution after cross-linking of precursors. Such fillers are used to achieve control of porosity regarding form, amount, size and distribution of pores in PDCs [SKKK2006]. Sacrificial fillers assist in the solution of the shrinkage issue because they function similarly to the passive ones – by reducing the volume fraction of the shrinking phase – and also because the resulting porosity increases the strain compliance of the system.

The use of fillers is an interesting approach, which increases even more the versatility of the PDC processing and the range of applications. Fig. 2.1.4 schematically summarizes the types of fillers and their respective effects in the PDC systems. Using one of these types or combinations thereof, novel materials with outstanding properties may be prepared with all the intrinsic advantages of PDC processing, as previously mentioned.

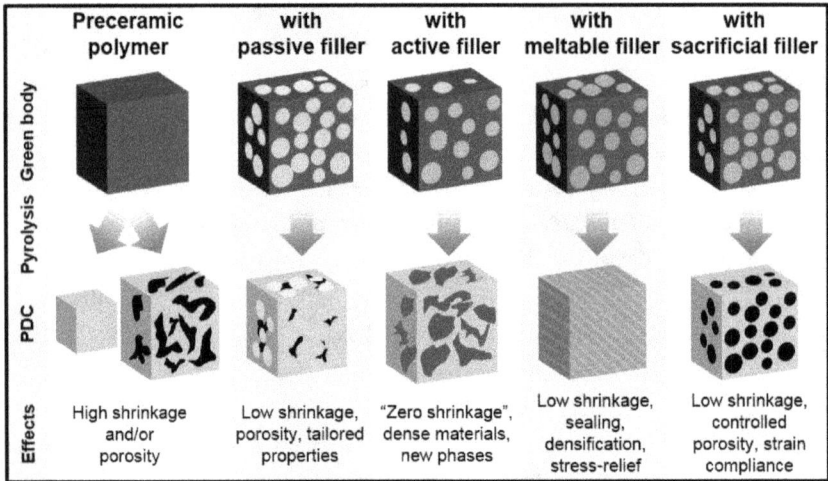

Fig. 2.1.4: Schematic representation of the types of fillers and their respective effects in the PDC systems.

## 2.2  Polymer-derived ceramic coatings

Ceramic coatings are the only known solution for several technical problems. Such coatings offer the possibility of applying usual structural materials under conditions beyond their limits. These structural materials, onto which the coating is deposited, is called substrate.

Ceramic coatings are used to increase the resistance of substrates against temperature [PaGJ2002], abrasion and wear [Berg2015], impact and perforation [GaNi2006], corrosion and oxidation at high temperatures [XSDB2015], and others. They might as well have other functions aside from protecting the substrate, e.g. to enhance surface area for catalytic applications at high temperatures [AgTs2000a, AgTs2000b], to enable biocompatibility [PWMM2016] or to modify electrical properties of surfaces [OlSB2001].

In general, the greatest drawback of ceramic coatings is the high costs associated with their processing, consequence of large energy consumption and/or long processing time. Due to simple procedures and relatively low temperatures (500-1200 °C), PDC processing arises as an interesting alternative to conventional methods for preparation of ceramic coatings, like physical vapor deposition (PVD), chemical vapor deposition (CVD), thermal spraying and sol-gel techniques [Ster1996, AeMe2004, ScTo2010].

Although also silicon-free polymers have been used to prepare ceramic coatings by PDC technique, only the state of the art of ceramic coatings derived from Si-based polymers – i.e. siloxanes, silanes, silazanes and derivatives thereof – was taken into consideration. The following literature review comprehends only reports about ceramic coatings prepared by cross-linking and ceramization of oligomeric or polymeric precursors deposited in liquid phase. The use of silicon-based polymers to prepare ceramic coatings by other methods, like CVD and sol-gel, and the coating of particles for the preparation of bulk ceramics were also considered beyond the scope of this work and are not reported herein. Reports on the use of preceramic polymer coatings to prepare micro- and nano-electromechanical systems (MEMS/NEMS) were also excluded from this review. Furthermore, a great variety of polymeric coatings have been developed from silicon-based compounds, including organic-inorganic hybrid coatings [KYWK1996, LCGP2005, HuXS2009, GPKM2014, CBBM2015, CBMS2015]. These systems where developed for applications, which did not required ceramization to obtain the desired properties. However, this chapter aims to elucidate the aspects of the preparation of ceramic coatings by PDC processing and, therefore, does not include polymeric coatings.

### 2.2.1 Processing of PDC coatings

PDC coatings may be prepared by a great variety of methods. The possibility of applying preceramic polymers in liquid phase (mostly in solution) enables the use of typical lacquer techniques for deposition of layers. These methods have advantages, such as simplicity and low cost in comparison to conventional ceramic coating deposition techniques. However, PDC processing of coatings requires additionally cross-linking and ceramization of the precursor. As previously discussed, precursor shrinkage during these processing steps limits the thickness of unfilled coatings to a few microns [GSGW2011]. By the use of fillers, preparation of thicker layers (above 20 µm) with tailored properties becomes possible.

It is important to mention that only general aspects of methods used for PDC processing of coatings are presented herein, whereas especial variations are not discussed. Furthermore, the suitability of the described methods to specific applications depends on several factors and must be evaluated for each case.

*Coating deposition techniques*

Among the great variety of deposition techniques for precursor-based coatings, four methods standout for their simplicity: dip coating, spray coating, spin coating and doctor blade method. In all these techniques deposition is carried out at room temperature and pressure, which is a great advantage for industrial applications. Independently of the selected technique, a liquid coating solution capable of wetting the substrate is always required to achieve good coating quality. A defect-free surface with suitable roughness must also be provided before coating deposition. For most applications, but especially for protective coatings, the surface must be completely covered to avoid localized damages and their propagation under the coatings. Hence, for rough surfaces, coating thickness is ideally at least as high as the roughness peaks, avoiding exposure of substrate. The surface must also be adequately clean and free from grease, dust, rust, or any kind of material, which could lead to weak adhesion, defects or failure of the coating. Occurrence of edge effects is also a factor in common to all techniques, whereas the extent of defects might vary. Usually, rounding or chamfering the edges of substrates avoid or at least reduce such defects. Moreover, each method has its particular characteristics, which will determine their suitability to different precursors and applications.

## Dip coating

Dip coating is the oldest deposition technique commercially applied [Brin2013] and the most common method for preparation of precursor-based coatings, together with spin coating technique. In this method, the substrate is immersed in the coating solution and removed vertically with a certain hoisting velocity, after a defined immersion time, to form the film [Brin2013]. Thus, deposition of the film occurs simultaneously in all the exposed faces of the substrate, which is, in some cases, a great advantage, especially for industrial applications. On the other hand, if only a specific area should be coated, masking of remaining areas is necessary.

The basic equipment for dip coating is simple and requires a small amount of operation parameters. An automatic control of immersion time and hoisting speed is sufficient to obtain excellent reproducibility. Losses of coating solution during processing are virtually zero and are mostly related to evaporation of solvents – to reduce evaporative losses, solution containers should have a small contact area with the atmosphere [OCCA1987]. However, if precision is required, this method is limited to simple geometries like sheets, plates, cylinders and tubes or slightly irregular shapes [YiMa2006], whereas deposition of coatings on substrates with complex shapes and cavities, although possible, is not uniform due to changes in flow of the liquid and a non-uniform drying process [APGA2004].

Coating thickness is dependent on hoisting velocity and on properties of the liquid (i.e. surface tension, viscosity, density and volatility of solvent). This dependency was discussed thoroughly for the first time by Landau and Levich in 1942 [LaLe1942]. According to their modelling – derived for Newtonian fluids without evaporative losses – and to experimental observations [MRRQ2011], coating thickness increases with increasing hoisting velocity, whereas the exact correlation between both is determined by properties of the liquid [YiMa2006].

Dip coating has poor performance near edges, leading to an inhomogeneous coating thickness. The extent of edge effects varies usually between 15 to 20 mm at top and bottom edges and between 8 and 15 mm at side edges [PuAe2004]. The effects increase with increasing hoisting velocity and are independent of substrate size [AEOR1997]. Another disadvantage of dip coating technique is the requirement of high volumes of coating solution in comparison to the volume that actually forms the coating. This factor is a drawback, considering that several precursors are air- and moisture-sensitive. Moreover, difficulties arise in dip coating processing during deposition of solutions with high viscosity and/or containing high amounts of fillers due to inhomogeneous deposition. Single dip coating procedure usually results in coatings with thickness up to a few microns, whereas repeating film deposition increases coating thickness. This requires, however, at least a cross-linking step between each repetition to avoid re-dissolution of the applied coating. Depending on precursor and desired coating thickness, even ceramization might be necessary between each repetition to avoid excessive overall shrinkage [PuAe2004].

Spin coating

Another common coating process is the spin coating technique, which has been applied on industrial scale since the 1950s. This method makes use of the centrifugal force acting on a spinning substrate to evenly distribute a liquid, which is usually deposited at the center of the surface, toward the edges [Birn2004].

As occurs in dip coating processing, coating thickness will be determined by the velocity and by properties of the liquid. However, in this case, an increase in angular velocity leads to thinner coatings [NoGL2005]. Indeed, coating thickness might as well be predicted or, at least, estimated in advance. The first model was described in 1958 by Emslie et al. [EmBP1958], although with several simplifications. Typical coating thicknesses range from a few nanometers to a few microns [LaRe1997]. In order to repeat the process and increase coating thickness, the issues regarding re-dissolution and shrinkage must be likewise addressed.

The basic equipment is simple and compact and offers excellent reproducibility. In comparison to dip coating, the volume of liquid required to apply coatings is small but waste of a high amount of coating solution is an important disadvantage of this technique [Kreb2009]. Additionally, restriction to single-face coating of flat or slightly curved substrates limits the range of applications, especially industrial ones. Notwithstanding, this technique is extensively applied by the electronic industry for coating of silicon wafers [NoGL2005, Kreb2009]. Difficulties arise, however, upon the use of coating solutions with high viscosity due to formation of defects [LNBC1993, LaRe1997, Kreb2009].

### Spray coating

The spray coating technique is probably the most common lacquer method, and has been applied since beginning of the 20th century [OCCA1987]. It is extensively used for industrial as well as for domestic coating applications. Deposition of films is carried out by atomization and transport of coating solution toward the substrate. Several methods for atomization have been developed, including airless, by air (carrier gas), supercritical, electrostatic and rotary atomization [Coel2012]. For processing of PDC coatings, the most common method is air atomization, in which the liquid is mixed with pressurized air and accelerated through a nozzle [IZYI2004]. For sensitive precursors, inert gases may replace compressed air as carrier gas. The deposition procedure, however, usually takes place in air, thus preventing the preparation of oxygen-free coatings and the use of extremely sensitive precursors.

Among all techniques, spray coating has the highest number of operation parameters, which include distance between nozzle and substrate, pressure of carrier gas, aperture of nozzle, shape of spray, angle of impact and velocity of the relative movement between nozzle and substrate. The high number of parameters is, at the same time, an advantage and a disadvantage. On the one hand, they provide wider flexibility regarding properties of the liquid (viscosity, size of filler particles, volatility of solvent), characteristics of deposited coatings (thickness and microstructure), and geometry of substrates [GRGH2009, Kreb2009]. On the other hand, they increase equipment's complexity and demand a great number of experiments to set optimum conditions.

The basic apparatus for spray coating with air atomization includes a gas source with pressure control, a feedstock container, a nozzle, and an exhaust system with filters, especially for solvent-borne and/or particle-containing coating formulations [OCCA1987]. The most common way to apply coatings by spraying is by means of spray guns and free-hand movements, by which reproducibility and uniformity of deposition will depend on the ability of the operator. To ensure reproducibility and uniformity of coatings, the use of mechanically or electronically controlled systems is necessary

[Coel2012]. The volume of liquid required to apply coatings is low [APGA2004], although a considerably high waste due to overspray might occur.

Doctor blade

Doctor blade is a technique, which has been employed since the 1940s for production of ceramic tapes (tape casting process). Nowadays, it is considered also an established coating method [HoBB1947, GRST1997]. In this technique, a liquid solution/suspension is fed from a reservoir and spread onto a surface by the relative movement between a blade and the substrate, which are separated by a defined gap. Two constructions are possible: moving blade with stationary substrate or moving substrate with stationary blade – the former is preferred for laboratories and discontinuous processing of small pieces, and the latter for continuous processing of long tapes [GRST1997]. In addition, the gap might be adjusted by either moving the blade or substrate. Feeding of liquid may be realized by means of a previously filled reservoir (discontinuous processing) or by continuous pumping into the reservoir [BeMS2004]. An automatic control of the velocity of blade/substrate movement provides very good reproducibility. For lab-scale coating application, simple frames with fixed gaps are commercially available, but reproducibility will depend on the ability of the user, since deposition is carried out by free-hand movements.

Parameters governing coating thickness are gap size, velocity of the movement and properties of the liquid [ChKY1987, BeMS2004, Kreb2009]. Typical coating thicknesses vary from a few microns to hundreds of microns [GRST1997]. If the coating solution is correctly dosed and evenly fed along the blade width, the method offers expressively low liquid consumption and waste [Kreb2009].

This technique is limited to flat or flexible surfaces, e.g. sheets and carrier tapes, and mostly to liquids with medium to high viscosity to avoid spontaneous flow of liquid under the blade, independently of the relative movement, especially if thicker coatings (larger gaps) are desired.

As described, each technique has its benefits and issues, being then suitable for different situations. In order to facilitate comparison, Fig. 2.2.1 summarizes the principle of each technique as well as their advantages and disadvantages.

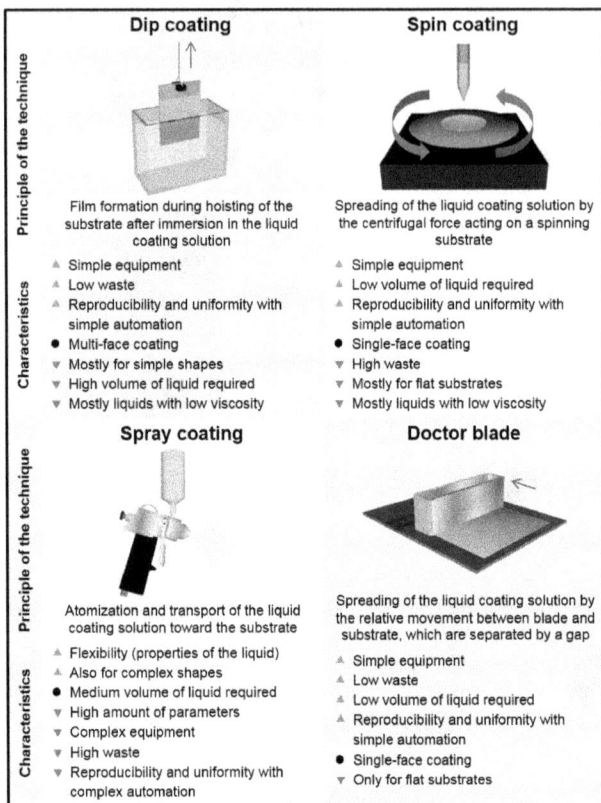

Fig. 2.2.1: Basic principle, advantages and disadvantages of the most common deposition techniques applied for the processing of PDC coatings.

*Cross-linking and ceramization processes*

Despite some common aspects in common with the processing of fibers, foams and bulk materials by PDC technique, some characteristics are exclusive of the processing of coatings. The most significant difference is the presence of a substrate, which may have properties completely distinct from those of the coating. However, both coating and substrate must undergo the same conditions in a conventional thermal treatment in furnace. Thus, the choice of parameters for cross-linking and ceramization of PDC coatings will strongly depend also on substrate's properties, such as melting point, thermal stability and oxidation resistance (for pyrolysis in air). This fact is the main motivation for the development of alternative methods, which can induce cross-linking and polymer-to-ceramic transformation without exposing substrates to damaging conditions. These methods enable application of more sensitive materials, even organic polymers, as substrates for ceramic coatings. Some examples of such techniques are laser pyrolysis [CMFV2001], chemically-induced conversion (moisture/catalyst) [MoSa2004, BDDE2005, Luka2007], ion irradiation [PiCT1996, PiCo1997a, PiCo1997b], and ellipsoidal mirror furnace with halogen lamp [GFHT2004]. Nevertheless, thermal conversion in furnace remains the most used method for preparation of PDC coatings on temperature-resistant substrates.

As previously discussed, mass loss and densification accompany the polymer-to-ceramic transition, and result in shrinkage of the material. Differently from free-standing parts, which may undergo an isotropic shrinkage, shrinkage of coatings is limited in two dimensions by adhesion with the substrate. Thus, shrinkage occurs freely only perpendicularly to the substrate surface (thickness shrinkage). This phenomenon, generally called constrained sintering, was substantially investigated in the late 1980s and early 1990s [ScGa1985, BoRa1985, ChRa1989, JaHu1990, JaHu1991, BoJa1993] and is a crucial aspect of the development of PDC coatings. Due to constrained shrinkage, there exists a critical coating thickness, up to which dense PDC coatings remain crack-free and stable. Above this critical thickness, intense cracking and spallation of coatings occur. The critical coating thickness of pure PDC coatings does not exceed a few microns [GSGW2011].

Additionally to intrinsic stresses generated by shrinkage of precursors upon ceramization, stresses arise during furnace pyrolysis and during application at high temperatures due to different coefficients of thermal expansion (CTE) of coating and substrate, as described by Eq. 2.2.1 [SoWa1999], which is a simplified mathematical expression to estimate thermal stresses in coatings:

$$\sigma_t = E_c(\alpha_s^l - \alpha_c^l)\Delta T \qquad \text{(Eq. 2.2.1)}$$

where $E_c$ is the Young's modulus of the coating, $\alpha_s^l$ and $\alpha_c^l$ are the linear CTEs of substrate and coating, respectively, $\Delta T$ is the variation of temperature, and $\sigma_t$ is the resulting thermal stress. How these thermal stresses affect coatings depends on their cohesive and adhesive properties. Coatings with high adhesion and cohesion have higher stress compliance. Coatings with high cohesion but weak adhesion will most likely fail at the interface with the substrate, whereas coatings with strong adhesion but low cohesion will generally fail within the coating.

According to Eq. 2.2.1, there are two approaches to reduce thermal stresses: to reduce the Young's modulus, e.g. by introducing porosity, or to reduce the CTE mismatch between coating and substrate. Porosity is a microstructural feature, which might not be desired, depending on the application – applications such as wear protection or environmental and diffusion barrier coatings require dense layers. Generally, the CTE of substrates are higher than that of coatings, due to the low CTE of silicon-based ceramics [BMSR2007]. Choosing substrates with low CTE and/or increasing the CTE of the coatings reduce the mismatch. Since the application requirements usually define beforehand which substrate must be used, tailoring of coating's CTE is the most common approach. This is realized by the introduction of fillers with high CTE, e.g. $ZrO_2$ or $CeO_2$, to increase the overall CTE of the coating. However, even the CTE of these ceramics (up to $15 \times 10^{-6}$ $K^{-1}$ [CaVS2004]) are still below the CTE of most metals, limiting the use of ceramic coatings onto metals. Furthermore, high temperatures intensify the effects of CTE mismatch. Consequently, the development of PDC coatings for high temperature applications becomes a great challenge and demands a considerable research effort.

The atmosphere also plays an important role on the processing of PDC coatings in furnace. The use of air as cross-linking/pyrolysis atmosphere is interesting, especially for industrial applications, due to lower processing costs. However, treatment in air of coatings deposited onto substrates with low oxidation resistance leads to formation of an oxide scale underneath the coating. These scales have in general a weaker adhesion than the coating itself, reducing the overall adhesion of the system. In more severe cases, damage of the substrate and degradation of its properties may also occur. Indeed, silicon-based ceramic coatings offer great oxidation protection and may enable cross-linking and pyrolysis of coatings on substrates with low oxidation resistance under oxidative atmosphere.

Another determining factor for the choice of atmosphere during thermal treatment is the intended composition of the coatings. While siloxanes already contain oxygen in the polymer structure, silazanes and silanes may be used to obtain oxygen-free ceramic coatings, e.g. $Si_3N_4$, SiCN, SiC. In these cases, the processing of coatings must be carried out under protective atmosphere, to avoid oxygen incorporation. Even oxygen-containing coatings may require pyrolysis under protective atmosphere, if presence of high amounts of carbon in the resulting ceramic is desired, e.g. to obtain SiCO ceramic coatings. Furthermore, the use of active fillers may also be a decisive factor to consider. Oxidation reactions occur faster and start generally at lower temperatures than nitride or carbide-forming reactions. Hence, depending on pyrolysis temperature, conversion of active fillers under nitrogen atmosphere may not be sufficient to overcome shrinkage of the precursor. Alternative methods, like plasma-assisted pyrolysis, have been developed to increase the reactivity of the nitrogen atmosphere, improving filler conversion [SGJM2016].

Therefore, the pyrolysis atmosphere is an important processing parameter and its effects on the properties of substrate and coating components must be considered in order to prepare PDC coatings by thermal treatment in furnace. In summary, the choice of atmosphere is usually determined by properties of the substrate and by the desired composition of the ceramic coating after pyrolysis.

### 2.2.2 State of the art of particle-filled PDC coatings

As previously mentioned, the use of fillers has mainly three purposes: solving the shrinkage problem, tailoring of microstructure or functionalization of coatings. These effects may be achieved by using ceramic, metallic, polymeric or meltable fillers. In fact, some systems even combine more than one type of filler to obtain coating systems for specific applications.

The first particle-filled PDC coating was presented by Labrousse et al., in 1993 [LNBC1993]. The authors developed SiC coatings derived from a polyorganosilane with varied filler contents. $Al_2O_3$ particles were used as passive fillers and pyrolysis was carried out in nitrogen atmosphere up to 900 °C with 5 h holding time. The authors concluded that the addition of filler particles enabled the increase of the critical coating thickness 10-fold (from 1 to ~10 μm). However, only pyrolysis temperatures up to 800 °C resulted in good coatings. Furthermore, the addition of filler reduced hardness and adhesion of the coatings.

Bill and Heimann [BiHe1996] investigated the development of transparent silazane-derived (NCP 200) protective coatings on carbon fiber-reinforced SiC composites. To reduce shrinkage, micro-sized silicon particles were added to the silazane. Coatings were prepared by dip coating followed by pyrolysis in nitrogen or argon at temperatures up to 1100 °C. Almost crack-free coatings with thickness of ~5 μm and pull-off adhesion of 14 MPa were obtained. These coatings significantly increased oxidation resistance of the substrates.

Also in 1996, Li and coworkers [LiKM1996] reported on polymer-derived SiCO layers for gas separation membranes. Multi-layered coatings were deposited onto porous alumina tubes by dip coating in solutions of polycarbosilane with different solvent amounts, to control porosity and infiltration of the substrate. Pyrolysis was realized under nitrogen or argon atmosphere at 350-950 °C for 1 h. Curing and pyrolysis of each layer was carried out before deposition of the next coating. The same procedures were executed with polycarbosilane solutions containing 5 wt% polystyrene (PS) particles as sacrificial filler. The authors concluded that pore volume, BET surface and, consequently, permeance of gases reached their maximum values after pyrolysis at 550 °C. Surface area and pore volume were slightly higher for PS-filled coatings, especially after pyrolysis at low temperatures. The authors continued their work with sacrificial fillers and a new study was published in 1997 [LiKM1997]. In this study, two more weight fractions of PS in polycarbosilane solutions were additionally investigated: 1 and 3 wt%. The two coating systems with lower PS content showed similar behavior, whereas coatings with 5 wt% resulted in higher permeance. All PS-filled coatings showed worse permselectivity than unfilled coating system.

In 2001, Colombo and coworkers [CMFV2001] presented an alternative ceramization technique to obtain SiC ceramic coatings from the polycarbosilane X9-6348: the laser pyrolysis. Polycarbosilane films with thickness of ~700 nm were deposited by spin coating on Si wafers and silica substrates. Laser pyrolysis was performed with a Nd:YAG laser in pulsed mode. In order to increase the absorption of laser radiation, a graphite layer was deposited at the surface of the green coatings. Alternatively, 5 vol% of graphite particles were added to the preceramic polymer. Both methods enhanced significantly the absorption of laser radiation, thus improving the pyrolysis process.

In 2006, Torrey et al. [TBHB2006] published a screening study to select suitable precursors and active fillers for coating applications. They investigated six different silsesquioxanes and an OH-modified polyhydridomethylsiloxane (PHMS) as precursor. Different metals (Al, Cr, Fe, Si and Ti) and one intermetallic compound (TiSi$_2$) were tested as possible active fillers. Thermogravimetric studies in air, argon and helium atmospheres have shown that the parameters of the thermal treatment are crucial to obtain filler conversion. A heating rate of 1 K min$^{-1}$ with dwell time of 1 h at 800 °C in air resulted in high conversion of all the investigated fillers.

In the same year, Suda and coworkers reported twice on porous coatings for gas separation applications. In the first paper [SYUF2006a], alumina tubes were coated with a polycarbosilane solution with and without PS particles by dip coating technique. Pyrolysis was carried out at 700 °C with 2 h dwell time, under argon atmosphere. The resulting coating thickness was below 1 μm. Gas permeation behavior was improved by adding filler particles. In the second paper [SYUF2006b], coatings of two different solutions of pure polycarbosilane (NIPUSI Type-S) were prepared by dip coating onto porous alumina tubes. One of the solutions contained the precursor as received and the other contained a pre-cross-linked precursor, obtained by reaction with a cross-linking agent and a catalyst for several days at 80 °C. Samples were pyrolyzed at different temperatures varying from 400 to 800 °C for 1 or 9 h, under argon flow. Additionally to pure polycarbosilane solutions, coating solutions were modified by adding PS (1 wt%). The authors determined that the pre-cross-linked precursor leads to smaller pore size. Furthermore, it was demonstrated that the sacrificial filler plays an important role in the microstructure evolution of the system and that reactions between precursor and gaseous products resulting from the decomposition of PS may occur.

Bakumov and colleagues [BGHS2007] investigated the introduction of silver nanoparticles in a polyorganosilazane-derived matrix (Ceraset™ VL20) to obtain coatings with antibacterial properties. Coatings were applied by dip and spin coating techniques on quartz glass and steel. Films were pyrolyzed under flowing ammonia and/or nitrogen for 60 min at temperatures ranging from 700 to 1000 °C. The authors have shown that the prepared coatings possess high antibacterial activity.

Henager et al. [HSBG2007] investigated the development of PDC coatings for nuclear energy systems. Although the focus of the paper was the development of joining layers, coatings composed of PHMS filled with SiC, Al or Al₂O₃ and combinations thereof were also developed. These coatings were deposited by dip coating onto SiC/SiC composites or stainless steel. Pyrolysis was carried out in air, nitrogen or argon atmosphere for 2 h at 1200 °C for coated ceramic substrates and at 800 °C for coated steel substrates. Porous coatings with thickness of above 100 μm were obtained on SiC/SiC substrates but no comments were made regarding coatings on steel substrates.

Elyassi and colleagues [ElST2007] also reported on the development of coatings for gas separation membranes using the polymer-derived ceramic route. Porous SiC tubes were firstly dip-coated with a solution of allyl-hydridopolycarbosilane filled with submicron SiC particles. Samples were heat-treated under argon atmosphere up to 750 °C. Four additional layers of pure polycarbosilane solution were alternately deposited and pyrolyzed following the same procedures. After pyrolysis of the fourth polycarbosilane layer, samples were treated at 450 °C for 2 h in air to oxidize eventual carbon residues. The overall coating thickness was ~2 μm. The applied procedures improved the reproducibility of properties and the hydrothermal stability in comparison to other PDC membranes prepared using the same precursor. In the following year, the same authors published a modification of the previous multilayered system [ElST2008]. In the new methodology, sacrificial PS layers were deposited before each new unfilled preceramic layer. The PS layers were likewise deposited by dip coating and were dried for 1 h at 100 °C. After pyrolysis of the fourth SiC layer and oxidation, thickness was larger than 7 μm. The authors affirmed that the PS layers not only block the infiltration of pores by subsequent precursor layers, but also participate in cross-linking reactions, as stated by Suda and colleagues [SYUF2006b, SYUF2006a] as well.

Torrey and Bordia published in 2007 and 2008 a series of papers on the development and properties of environmental barrier coating systems using the polyorganosiloxane PHMS with active fillers, based on results obtained in a screening study published in 2006 [TBHB2006]. In the first paper [ToBo2007], titanium disilicide was used as filler and coatings were applied onto stainless steel by dip coating. Pyrolysis was carried out in dry air at temperatures varying from 200 to 800 °C. Coatings with thickness of ~18 μm were obtained. They observed that, despite the use of the active filler, about 10 vol% porosity remains in the coatings after pyrolysis at 800 °C. Also some carbon content is still present after pyrolysis. The second paper [ToBo2008a] reports about mechanical properties of the same system. Hardness, elastic modulus, tensile strength and ultimate shear strength of the interface of coatings were investigated. In the last paper [ToBo2008b], the authors report once again about particle-filled PHMS coating systems, this time with different active fillers: Cr, Fe, Ti and $TiSi_2$. Coatings were applied by dip coating onto stainless steel substrates and the pyrolysis procedure was the same described in the previous paper. Coatings with thickness of 15-20 μm were obtained. However, only the system with $TiSi_2$ performed satisfactory, forming well adherent, crack-free coatings. The authors reported that the use of submicron powder of active filler increased conversion into oxide. Furthermore, formation of a diffusion zone at the interface substrate/coating was observed.

Also in 2008, Corral and Loehman [CoLo2008] investigated the development of ultra-high-temperature ceramic coatings for aerospace applications. In their study, C/C composites were coated with allylhydridopolycarbosilane (SMP-10) filled with SiC particles by dip coating under reduced pressure. Pyrolysis was conducted under argon atmosphere up to 1100 °C and 1 h holding time. These procedures were repeated up to three times to increase coating thickness. A second coating of $ZrB_2$ was deposited onto the polymer-derived SiC coating. A good protective performance against oxidation was observed upon short-time exposure to high temperatures.

Pavese et al. [PFBO2008] investigated the preparation of particle-filled polycarbosilane coatings for protection of C/SiC CMCs for aerospace applications. C/SiC composites prepared by PIP (precursor infiltration and pyrolysis) were brushed with a coating suspension containing allylhydridopolycarbosilane and $HfB_2$ particles. Pyrolysis was conducted at 1000 °C under inert atmosphere. The process was repeated two more times to obtain homogeneous coatings with thickness of 35-50 μm. Density and Young's modulus were slightly increased, whereas the flexural strength was not affected. Despite similar mass losses when heated to temperatures of 1600 °C, coated C/SiC substrates underwent less strength loss in comparison to uncoated samples, which was attributed to different damage mechanisms.

In 2009, Günthner and coworkers [GKKM2009] published a paper about silazane-based protective coatings. Hexagonal boron nitride (hBN) was used as passive filler in PHPS (perhydropolysilazane, NN120-20) coatings to obtain dense layers with ~12 μm of thickness after pyrolysis in air up to 800 °C. Coatings were deposited onto mild and stainless steel sheets. The authors calculated oxidation rates up to two orders of magnitude lower for coated substrates in comparison to the uncoated counterparts. Performance was also better in comparison to unfilled PHPS coatings, which was attributed to the higher coating thickness of the particle-filled system.

In the same year, Kraus et al. [KGKM2009] published a paper about properties of an ABSE (ammonolysed bis(dichloromethylsilyl)ethane) coating system with cubic boron nitride (cBN) as filler. Coatings with thickness up to ~15 μm on stainless and mild steel substrates were obtained after deposition by dip coating and pyrolysis at 800 °C for 1 h in air. The use of the filler reduced coating shrinkage and enabled the preparation of thicker coatings compared to unfilled ABSE layers. Furthermore, coated samples showed improved oxidation resistance compared to uncoated metal sheets.

Also in 2009, Stern et al. [StHS2009] published their study on micropatterned PDC coatings by phase separation technique using methanol as non-solvent. Coatings were prepared from a mixture of polymethylphenylvinylsiloxane (Silres® H62C), polymethylsiloxane (Silres® MK) and methyltriethoxysilane (MTES) with SiC as particulate filler. Coatings were applied onto ceramic tapes, steel and glass by dip or spin coating. After pyrolysis at 1000 °C in nitrogen atmosphere for 4 h, highly porous coatings (more than 40% porosity) with thickness of ~28 μm were obtained.

In 2010, Kappa and coworkers studied the application of organosiloxane-based coatings – polymethoxymethylsiloxane (PMMS) and hydroxy-terminated polydimethylsiloxane (PDMS) – on steel substrates. Unfilled coatings were compared to coatings containing 5 vol% of SiC, $Si_3N_4$ or $Al_2O_3$. Coatings were prepared by dip coating and were pyrolyzed at 1000 °C in argon atmosphere. The formation of a 1.5 μm thick diffusion zone constituted of chromium carbide and nickel-chromium spinel was observed at the interface with the substrate. Furthermore, the use of SiC and $Si_3N_4$ as fillers resulted in higher mass losses compared to unfilled precursor, whereas $Al_2O_3$ led to lower mass losses during pyrolysis. The use of $Si_3N_4$ as filler also resulted in patterned surfaces, attributed to de-wetting and de-mixing phenomena.

Liu and colleagues [LZLC2010] also investigated PDC coatings as environmental barriers. In their study, C/SiC composites were coated with a polysiloxane (Elastosil® RT 601) solution filled with BSAS – barium-strontium aluminosilicate – by brushing. Pyrolysis was carried out at 1350 °C under argon atmosphere. Layers with thickness of about 20 μm were obtained. Protection against oxidation and water vapor attack was provided by the coatings upon exposure to temperatures up to 1250 °C.

Yang and colleagues [YYZS2011] compared pure SiC coatings prepared from a self-synthesized polycarbosilane with layers prepared using the same precursor filled with aluminum particles. In their study, Fecralloy was used as substrate and coatings were deposited by dip coating. Pyrolysis was realized up to 1000 °C. Crack-free coatings with thickness up to 2 μm were obtained using the pure polycarbosilane, whereas thicknesses up to 35 μm were obtained using the active filler approach. The authors observed the formation of an Al/Fe mixed zone at the interface. Moreover, the hardness of free-standing coating material was higher than that of the coatings, which was attributed to synergistic effects between substrate and coating.

Also in 2011, Günthner and Wang et al. published a two-part study on particle-filled PDC coatings. In the first part, Günthner et al. [GSGW2011] reported on the development of thick EBCs based on the combination of a silazane with passive and glass fillers. A ~1 µm PHPS (NN120-20) bond-coat was applied to protect the mild steel substrate against oxidation during pyrolysis and to enhance adhesion of the top-coat. Different top-coat compositions were tested, whereas the best results were obtained with polyorganosilazane HTT 1800 filled with zirconium dioxide and two types of glass with different softening temperatures. Coatings were applied by dip coating and were pyrolyzed at temperatures up to 800 °C in air for 1 h, resulting in layers with thickness above 70 µm. Moreover, the good adhesion of the coatings was attributed to the surface pre-treatment by blasting with glass beads and to the bond-coat. Developed coatings were able to protect mild steel substrates against oxidation even during prolonged exposure to high temperatures in air. In the second part, Wang et al. [WGMB2011] studied properties of a PHMS-derived SiCO coating filled with zirconium disilicide with and without a PHPS-derived (NN120-20) SiON bond-coat. Properties of PHPS-derived SiON coatings alone were additionally investigated. Coatings were deposited onto nickel-based superalloy substrates by dip coating. All coatings were pyrolyzed in air at 800 °C for 2 h. Composite coatings with thickness of 20-25 µm were prepared. The most efficient protective coating against oxidation was determined to be the double-layered system.

Woiton et al. [WHLS2011] investigated properties of coatings obtained from variations of the PDC coating system published by Stern et al. in 2009 [StHS2009]. The amount of each of the components – polymethylphenylvinylsiloxane (Silres® H62C), polymethylsiloxane (Silres® MK) and MTES – was varied and addition of SiC as particulate filler was investigated. Coatings were applied onto $Al_2O_3$ tapes by dip coating technique. Samples were pyrolyzed in nitrogen atmosphere up to 1100 °C with holding time of 2 h. The authors found that the filler particles changed the microstructure of the coatings significantly.

A year later, Schütz et al. [SGMG2012] published a study, in which the coating systems with polysilazane and glass fillers developed by Günthner et al. [GSGW2011] were applied and further characterized. Coatings with thickness up to 100 µm were obtained after deposition by dip coating onto mild and stainless steel substrates and pyrolysis in air at 700 °C for 1 h. A PHPS (NN120-20) bond-coat was previously applied by dip coating and pyrolyzed in air at 700 °C for 1 h. The thermal conductivity of these coatings was measured by laser-flash and amounted to 0.94 W m$^{-1}$ K$^{-1}$ at room temperature and 0.65 W m$^{-1}$K$^{-1}$ at 600 °C. Moreover, the growth of needle-like structures at the interface coating/substrate was observed. The investigated coatings performed well upon abrasion at high temperatures.

In the same year, Wang et al. [WYLF2012] also reported on silazane-based coatings with glass fillers to protect C/C-SiC composite materials against oxidation. Triple-layered coatings were prepared using a non-commercial polysilazane filled with one of three types of glass: a borosilicate, a bismuth-based and a lead-based glass. Coatings were prepared by brushing a layer onto the composites, followed by curing with an infrared lamp and application of the next layer onto it. Density, porosity, composition and microstructure were investigated and oxidation and thermal shock tests up to 700 °C were carried out. The coatings were able to protect the substrate against oxidation under the applied conditions.

Wang and colleagues [WTQG2012], also in 2012, reported on the development of PDC coatings with high infrared emissivity. Coatings composed of PHMS, Al particles as active filler and SiC-based powders as functional filler to enhance infrared emissivity were deposited onto stainless steel by dip coating. Pyrolysis was realized in air at 600 °C or 800 °C for 2 h. Coatings with thickness of ~15 µm were characterized regarding composition, microstructure, emissivity in infrared range and thermal shock resistance. Infrared emissivity was significantly improved by the coatings.

Jung et al. [JSLF2012] reported on the development of polyorganosilazane (Ceraset™ Polysilazane 20) layers filled with SiC, TiB$_2$ or Al particles to prepare resistance-temperature detectors (RTD). Layers were applied by screen printing on quartz wafers followed by pyrolysis under nitrogen atmosphere at 1100 °C with holding time of 3 h. The active fillers TiB$_2$ and Al performed better and enabled the preparation of coatings with nearly zero shrinkage.

Liu and coworkers [LZYC2012] reported on the preparation of environmental barrier coatings composed of SiCN and Sc$_2$Si$_2$O$_7$ for C/SiC CMCs. The authors prepared coatings using a polysilazane as SiCN precursor, scandium disilicate – self-synthesized by sol-gel – as passive filler, and Li$_2$CO$_3$ was used as sintering aid. Coatings were applied onto C/SiC composites with a brush and pyrolysis was conducted in a multistep process up to 1250 °C under argon atmosphere. Coatings with thickness of ~30 µm were obtained after pyrolysis. The coatings were able to protect the substrate against corrosion by water vapor at high temperature. One year later, the authors published another paper [LZHY2013] on the use of the active filler approach to prepare polymer-derived yttrium silicate EBCs. Yttrium silicates were obtained by pyrolysis of polysiloxanes (ELASTOSIL®

RT 601) with different amounts of $Y_2O_3$ filler particles and $Li_2CO_3$ as sintering aid. Coatings were deposited onto C/SiC composites and pyrolysis was carried out at 1400 °C for 5 h under argon atmosphere, resulting in dense, crack-free coatings with thickness of ~20 μm. The developed coatings showed promising protective properties against water vapor corrosion.

Also in 2013, Tian and colleagues [TWLG2013] continued the work of Wang et al. [WTQG2012] and reported on the development of coatings with high infrared emissivity by PDC processing using active and passive functional fillers. Coatings of PHMS with Al particles as active filler and $Cr_2O_3$ as functional component to increase emissivity were applied by dip coating onto stainless steel substrates. Pyrolysis was carried out at 600 °C or 800 °C in air for 2 h, resulting in coatings with thickness of 10-20 μm. Like the SiC-filled coatings investigated by Wang et al., chromia-filled coatings were able to increase significantly infrared emissivity.

A study published by Wang et al. [WUTF2014] in 2014 investigated once more properties of the $PHMS/TiSi_2$ coating system developed and investigated by Torrey and Bordia [ToBo2007, ToBo2008b, ToBo2008a], this time with focus on corrosion and oxidation protection. The procedures for preparation of the coatings were the same used by Torrey and Bordia – dip coating onto stainless steel 316 and pyrolysis in air at 800 °C for 2 h. Coatings with thickness of 12-15 μm were obtained. To increase coating thickness and close open pores, deposition and pyrolysis procedures were repeated up to four times, resulting in coatings with thickness up to 48 μm. Pure PHPS (NN120-20) coatings were also prepared and investigated. The coatings were able to improve corrosion/oxidation resistance regarding acidic corrosion in sulfuric acid and cyclic and static oxidation. The composite coatings performed better than PHPS-derived ones due to larger thickness.

Also in 2014, Xiao et al. [XZZL2014] reported on properties of thin PDC coatings obtained by pyrolysis of a self-synthesized polysilazane with aluminum particles. Sheets of stainless steel were dip-coated and subjected to a multi-step pyrolysis in nitrogen atmosphere up to 800 °C. Coatings with thickness up to ~5 μm were obtained. According to the authors, crack-free and well adherent coatings could be obtained with filler amounts up to 30 vol%. However, hardness and Young's modulus diminish with increasing amount of filler.

Tang and colleagues [TFHZ2015] investigated the development of $ZrSiO_4$-SiBCN(O) coatings for oxidation protection of SiC substrates. Coatings were prepared using a self-synthesized polyborosilazane as preceramic polymer and $ZrB_2$ as filler. Pyrolysis was conducted in nitrogen atmosphere at 1300 °C. Coatings with thickness of ~8 μm were obtained. Formation of zirconium silicate was observed. Upon oxidation at

1500 °C, a borosilicate passivating layer forms at the surface, enhancing oxidation resistance and the protective character of the system.

In a recent study, Biasetto and coworkers [BEBC2016] prepared coatings from polymethylsiloxanes with nano and micro-sized $TiO_2$, $CaCO_3$ and sphene ($CaTiAlO_3$) particles to obtain sphene coatings on titanium for biomedical and dental applications. Coatings were deposited by spray coating on titanium plates and pyrolysis was conducted in air at 950 °C for 3 h. Crack-free coatings with thickness up to ~100 µm and low porosity were obtained. Furthermore, the good adhesion of the coatings was attributed to the formation of a diffusion zone at the interface with the substrate.

Another recent publication, of Barroso et al. [BKDS2016], also reports on the development of PDC coatings with filler particles. In the first part, polyorganosiloxanes PMMS and PDMS were combined to prepare coatings on steel substrates. Micro- and nano-sized spheres were prepared from the same precursors and used as fillers to obtain structured surfaces. Coatings were applied by dip coating and pyrolysis was carried out in argon atmosphere at temperatures up to 1000 °C. In the second part, silazane-derived coatings with hexagonal boron nitride filler particles were investigated regarding the applications as diffusion barrier coatings or as wear protection. Diffusion barrier coatings were prepared using the organosilazane Durazane™ 1800 as precursor. Coatings were applied by spray coating onto graphite plates and pyrolysis was conducted in nitrogen atmosphere at 1000 °C for 1 h. Coatings with thickness of ~25 µm were obtained. These coatings are able to reduce the diffusion of carbon from the graphite dies into $Si_3N_4$ ceramic products during a real industrial hot press sintering procedure. The second silazane/hBN coating system was prepared using ABSE as precursor. This system was applied by dip coating onto steel substrates and pyrolyzed in air at 800 °C for 1 h. Coatings with thickness of 8-10 µm were prepared. Performance of this coating system and of cBN-filled coatings developed a few years earlier by Kraus et al. [KGKM2009] was investigated with respect to abrasion resistance and tribological behavior. Due to low hardness, the hBN-filled system did not performed well, whereas the cBN-filled coatings showed improved resistance.

## 2.3 Thermal barrier coatings

Gas turbines have found extensive use in aerospace and energy sectors and the minimum requirements of efficiency and performance of these turbines are constantly increasing. However, in order to enhance their efficiency, they must operate at higher temperatures. Degradation of mechanical and thermomechanical properties of metallic turbine parts has been the limiting factor preventing further temperature increase. The introduction of blades with textured microstructures and later of blades formed by single crystals, together with internal cooling, have enable a significant temperature increase [ClPh2005]. With the development of thermal barrier coatings composed of refractory materials with low thermal conductivity, a substantial progress in efficiency increase of gas turbines was realized. Thermal barrier coatings were firstly applied on vane platforms of gas turbine engines in the 1970s and their use has been extended to other turbine parts ever since [Mill1997]. The main function of the TBC in a gas turbine is to reduce the rate of thermal transport from the hot gas stream to the substrate. Combined with internal cooling, TBCs with a coating thickness of 100-500 μm enable a reduction of up to 300 °C in the temperature of the substrate, currently produced from nickel- or cobalt-based superalloys [PaGJ2002, KuKa2016], enabling operation of turbines with gas temperatures even higher than the melting temperature of the alloying elements in the substrate (~1300 °C) [PaGJ2002, ClPh2005, KuKa2016]. From another perspective, a reduced substrate temperature increases lifespan and durability of the metal parts, making the use of TBCs interesting even for lower temperatures.

A TBC system is generally composed of three layers applied on metallic substrates (Fig. 2.3.1). The first layer is called bond-coat. This 75-150 μm thick coating consists mainly of metallic alloys – e.g. NiCrAlY, NiCoCrAlY or aluminides of Ni and Pt – which possess high oxidation resistance [PaGJ2002]. The main function of this layer is to prevent excessive oxidation and corrosion of the substrate, improving compatibility between top-coat and substrate. Deposition techniques include plasma spraying, EB-PVD, electroplating and CVD [PaGJ2002].

Despite the temperature reduction realized by combining TBC and internal cooling, the temperature of the bond-coat under normal operation may exceed 700 °C, resulting in oxidation of the metallic layer. Hence, the second layer in a TBC system is the thermally grown oxide (TGO) layer, spontaneously formed on the metallic bond-coat. This layer has thickness ranging from 1 to 10 μm and its composition is determined by the composition of the bond-coat, whereas formation of a slow-growing and defect-free $\alpha$-$Al_2O_3$ layer – $Al_2O_3$ has the lowest oxygen diffusivity among the common oxides [ClPh2005] – is the most desired TGO [PaGJ2002]. The growth of the TGO layer may occur by inward diffusion of oxygen through the TGO, by diffusion of Al from the bond-coat toward the surface or by a combination of both. Due to the very low oxygen diffusivity, uniform $\alpha$-$Al_2O_3$ layers are able to deaccelerate growth rate of the TGO layer [PaGJ2002]. As will be discussed hereafter, growth of the TGO layer is one of the main causes of TBC failure and must, therefore, be well controlled. The composition of the bond-coat may be further tailored in order to maximize adhesion of the TGO. While elements like S, Ti and Ta diminish adhesion, others like Si and Hf increase it [PaGJ2002].

The last layer is the one responsible for the thermal barrier functionality and must have, therefore, a low thermal conductivity. Another basic requirement is a high thermal expansion to relieve stresses in the coatings caused by the mismatch between the CTE of the substrate (CTE = $14\times10^{-6}$ $K^{-1}$) and that of the coating [PaGJ2002]. The porosity/cracking pattern of this layer influences both thermal conductivity and strain resistance. While pores and cracks parallel to the direction of thermal transport contribute to a reduction of thermal conductivity toward the substrate, a perpendicular orientation is rather beneficial for a better strain resistance [PaGJ2002]. It was estimated that for each 25 µm thickness of a typical top-coat, substrate temperature may be reduced by 4-9 °C, difference which is further enhanced by internal cooling of the parts [KuKa2016]. The most common methods for preparation of top-coats are APS and EB-PVD. The deposition method determines the microstructure and the failure mechanisms occurring in the coatings. Therefore, deposition methods and their features are discussed further in the following sections.

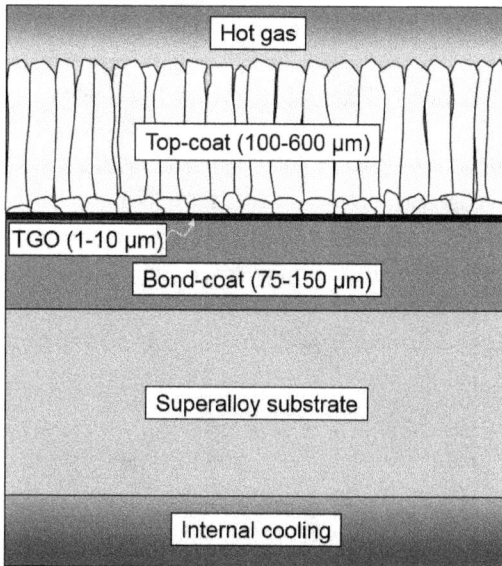

Fig. 2.3.1: Basic structure of a thermal barrier coating system. Based on [PaGJ2002].

### 2.3.1 Materials of interest for the top-coat

The primary requirements for a top-coat material are properties, such as high melting point, low thermal conductivity, CTE as close to that of the substrate and no phase transformations between room temperature and operation temperature [ClPh2005]. Other properties like chemical inertness, good adherence to the metallic bond-coat and low sintering rate are also important characteristics to be considered [CaVS2004].

Based solely on requirements of low thermal conductivity and high CTE, zirconium dioxide would be a good candidate for TBC applications. However, zirconia undergoes two phase transitions, depending on the temperature – at about 1170 °C, the monoclinic structure transforms into tetragonal and, at about 2730 °C, the tetragonal phase transforms into cubic zirconia [CGVC2009]. Especially the transformation from monoclinic into tetragonal has significant influence on performance of thermal barrier coatings, as it occurs in the range of the application temperatures. This transformation causes a volume retraction of 4-6%, which is reversed upon cooling down, when the structure transforms back into the monoclinic allotrope [BeMv1999]. Volume changes and shear strain caused by phase transitions lead to high internal stresses and, consequently, to coating failure. However, the addition of suitable oxides, like $Y_2O_3$, $Al_2O_3$, $La_2O_3$, $CeO_2$, MgO and CaO, can stabilize the high temperature phases of zirconia, generating metastable phases at low temperatures [PaGJ2002, ClPh2005, KeDe2008]. Depending on the amount and type of stabilizer, the resulting material may be classified as fully stabilized zirconia (FSZ), if cubic structure is the only phase; partially stabilized zirconia (PSZ), if the cubic phase contains nanosized particles with monoclinic or tetragonal structures; or tetragonal-stabilized zirconia (TSZ), if the material is composed mainly of the tetragonal allotrope [CGVC2009]. Yttria-stabilized zirconia (YSZ) – especially the variety containing 4-4.5 mol% yttria, with metastable tetragonal structure [ClPh2005, PPWC2012] – is the most common material for TBC applications. Additionally to phase stability, it has low and, unlike most crystalline oxides, temperature-independent thermal conductivity (2-3 W m$^{-1}$ K$^{-1}$). Moreover, it possess high coefficient of thermal expansion (~11.5×10$^{-6}$ K$^{-1}$) and high melting temperature (2680 °C) [VJSM2010].

For current requirements of temperature stability and durability, YSZ top-coats perform well. However, in order to further enhance efficiency, an increase in the gas inlet temperature is necessary, requiring performance and durability beyond the limits of current TBC systems with YSZ top-coats [LPPV2003, VJSM2010]. Indeed, upon long exposure to temperatures beyond 1200 °C, the metastable tetragonal phase of YSZ decomposes into high and low yttria phases, whereas the low-yttria fraction undergoes tetragonal-to-monoclinic transformation upon cooling, leading to coating failure [BrTa1991, TaBM1992]. Sintering is another issue related to long-term exposure to high temperatures, which leads to performance loss and eventually failure [CiGC2009]. Hence, substitutes for YSZ must be found and a significant effort has been dedicated to research and development of new top-coat materials.

Three classes stand out among the candidate materials. The first is that of pyrochlores, defined by a general composition $A_2B_2O_7$ – especially those with the element B = Zr, Hf or Ce – due to their low thermal conductivity (1-2 W m$^{-1}$ K$^{-1}$) [SSKM1998, VCTB2000, LPPV2003, MGXC2006, VJSM2010, PPWC2012]. The second class is that of hexaaluminates, with a general composition (La,Nd)MAl$_{11}$O$_{19}$ and M = Mg, Mn-Zn, Cr or Sm, which combine high melting point, high CTE, low thermal conductivity, low sintering rates and phase stability up to 1800 °C [ISEA1999, FrGS2001, VJSM2010]. Also perovskites, with basic composition ABO$_3$ – especially zirconates (B = Zr) – is another class of materials of interest for top-coats. The CTE of zirconates perovskites is generally lower, and the thermal conductivity mostly similar to that of YSZ [ClPh2005, MJMP2008, VJSM2010]. However, the crystalline structure of perovskites enables incorporation of a broad variety of ions, increasing the potential of these materials for TBC applications. Indeed, much research effort is still required to establish other materials for application as top-coat in TBC systems and thus, YSZ remains the standard material for TBC applications.

### 2.3.2 Thermal properties and failure mechanisms

Processing of TBC top-coats may be carried out by several methods, mostly based on CVD, PVD and thermal spraying techniques. However, two methods stand out and are the most commonly used for deposition of top-coats: APS and EB-PVD. Depending on the technique employed, top-coats of the same materials may have completely different microstructures and properties.

Not only the thermal properties but also the durability of a TBC is strongly dependent on microstructure. In general, failure of a TBC is caused by constrained growth of the TGO layer, causing intense, temperature-independent compressive stresses (up to 1 GPa [SSBP2000]). While the top-coat has cracks and porosity to relieve stresses, the TGO is ideally dense, resulting in higher residual stresses [PaGJ2002]. The mismatch between CTEs of metallic substrate/bond-coat/TGO composite and of ceramic top-coat leads to additional compressive stresses upon cooling, which may reach up to ~7 GPa [LiCl1996]. Furthermore, an extensive growth of the TGO layer may lead to a scarcity of Al in the bond-coat, resulting in the formation of oxides other than $\alpha$-Al$_2$O$_3$, consequently causing additional stresses in the TGO layer and increasing oxygen diffusivity, thus accelerating further TGO growth [Stra1985]. Segregation of undesirable elements like S, Ti and Ta at the interfaces is also prejudicial and may lead to reduction of adhesion and toughness [PaGJ2002]. The effect of these stresses within the coatings and at the respective interfaces also depends on the microstructure of deposited coatings. However, the microstructure may change during application due to sintering. A reduction of voids fraction and a partial or complete elimination of crack patterns accelerates failure due to increased Young's modulus, increasing thermal stresses and reducing strain tolerance [LPPV2003]. Furthermore, elimination of porosity and cracks increases thermal conductivity, leading to an increase in substrate temperature, resulting in faster TGO growth and creep, hence, accelerating failure [ZhMi1998].

Due to the dependence of properties of the coatings and their respective failure mechanisms with the deposition method, these topics are discussed separately for each of the two most commonly used techniques – APS and EB-PVD. Although TBC top-coats may be prepared from different materials, the discussions refer to YSZ top-coats.

*APS top-coats*

In this technique, the feedstock is melted in a plasma flame and accelerated toward the substrate, where it resolidifies, forming the so-called splats (flattened particles) [HeSa1996]. Splats – typically 1-5 μm thick with a diameter of 200-300 μm – form parallel to the surface, hence splat boundaries, and thus also cracks related to these boundaries, are oriented in the same direction, reducing thermal conductivity [PaGJ2002]. Top-coats deposited by this technique possess porosity in the range of 15 to 25 vol%, which contributes to the strain resistance and reduces the overall thermal conductivity [PaGJ2002]. Due to the obtained microstructure, APS top-coats have thermal conductivity in the range of 0.8 to 1.7 W m$^{-1}$ K$^{-1}$ [PFSK1997, PaGJ2002]. A typical top-coat prepared by APS has thickness around 300 μm but coatings with thicknesses of more than 600 μm may be produced [PaGJ2002].

The APS technique requires a rough (ideally undulated) bond-coat surface in order to obtain top-coats with sufficient adhesion – i.e. the dominant adhesion mechanism is mechanical interlocking [PFSK1997]. However, this roughness increases the surface area of the bond-coat subjected to oxidation [Stra1985] and lead to out-of-plane stresses upon growth of the TGO layer and thermal cycling [PaGJ2002]. The growth of the TGO layer results in stresses at the interfaces with bond-coat and top-coat, which are tensile at crests and compressive at troughs [GoCl1998, HBFL1999]. The tensile stresses lead to adhesive failure at crests of the bond-coat/TGO and TGO/top-coat interfaces. Moreover, tensile stresses also cause cracking parallel to the substrate within the top-coat at the vicinity of crests [SPJG2003]. However, beyond a certain TGO layer thickness, the CTE of the bond-coat/TGO assembly becomes lower than the CTE of the top-coat, transforming the tensile stresses in crests into compressive, resulting in hoop tension and cracking parallel to the substrate within the top-coat region between crests [HsFu2000, SPJG2003]. These cracks then propagate through the TGO and coalesce with cracks resulted from adhesive failure at crests of the bond-coat/TGO interface [RaEv2000].

The APS technique is a somewhat versatile technique with lower production costs in comparison to other established techniques for deposition of ceramic coatings. However, due to the described failure mechanisms, APS TBCs are in general more sensitive to thermal cycling than the EB-PVD counterparts. Indeed, APS TBCs have been successfully applied mostly in regions of aircraft engines subjected to less harsh conditions – combustors, fuel vaporizers, after-burner holders, stator vanes – and more broadly in industrial gas turbines due to lower operating temperatures and reduced thermal cycles [PaGJ2002].

The limitation of the APS technique relates to the expensive and complex equipment, which, due to its size, is not able to reach small hidden areas, like the inner face of long pipes with small diameters.

*EB-PVD top-coats*

EB-PVD consists in the bombardment and evaporation of a target by an electron beam inside a vacuum chamber. The evaporated material is then deposited onto the substrate, where it resolidifies [Zhan2011]. In contrast to APS, EB-PVD provides coatings with better durability when applied onto smooth surfaces, as adhesion occurs mainly by chemical mechanisms [Stra1985, PFSK1997]. However, the thickness of EB-PVD coatings is in most cases lower than that of APS top-coats (~125 μm) [PaGJ2002]. The microstructure of EB-PVD top-coats is in general characterized by two zones. At and in the immediate vicinity of the interface with the TGO layer, the top-coat is dense and composed of equiaxed grains with 0.5-1 μm [PaGJ2002]. This initial layer – with thickness below ~2 μm to prevent excessive compressive stress due to CTE mismatch – is obtained by deposition of YSZ with poor oxygen conditions and is responsible for the chemical adhesion of EB-PVD top-coats [Stra1985]. By changing the atmosphere to oxygen-rich, nanoporous columnar grains with diameter of 2-10 μm and separated by void channels grow out of the dense region [Stra1985, PaGJ2002]. While the orientation of cracks and pores parallel to the substrate, characteristic of APS top-coats, contributes to a low thermal conductivity, the columnar structure of EB-PVD increases the strain tolerance of top-coats. Upon thermal cycling, stresses caused by the higher CTE of the metallic substrate/bond-coat/TGO assembly in comparison to that of the top-coat are relieved by free expansion or contraction of the gaps between columns [Stra1985]. On the other hand, this type of microstructure contributes in a more limited extent to a reduction of the thermal conductivity in comparison to dense YSZ ceramics, resulting in typical values around 1.5 W m$^{-1}$ K$^{-1}$ [PFSK1997].

Due to the increased strain resistance of the top-coats, failures in EB-PVD TBCs occur mainly at interfaces. One of the possible failure mechanisms is the adhesive failure at the interface bond-coat/TGO layer on the top of imperfections (ridges) of the bond-coat surface [GJVM1999, VaGJ2000] – similarly to failure on the undulation crests at the same interface in APS TBCs. Another mechanism is characterized by an adhesive failure at the TGO/top-coat interface caused by a deformation of the bond-coat/TGO surface. Possible causes for this mechanism are the roughening of the TGO due to creep, fast oxidation of other metals than aluminium due to TGO cracking, and formation of depressions on the bond-coat [CJBG1998, GJVM1999, MuEv2000, HeEH2000, EvHH2001, SKJG2001]. Furthermore, even EB-PVD coatings with defect-free interfaces may fail due to buckling, caused by compressive stresses within the coatings [EvHu1984, VaGJ2000].

Despite the higher durability of EB-PVD TBCs, reduced thickness and higher processing costs when compared to APS make this technique suitable mostly for coating of parts subjected to harsher conditions, such as turbine vanes and blades [PaGJ2002, VJSM2010]. Furthermore, the requirement of a vacuum chamber large enough to contain the part to be coated limits the size of substrates. Also hidden surfaces are an issue due to shadowing effects, which may lead to changes in microstructure or even prevent adequate coating of these hidden areas.

## 2.4    Automotive exhaust systems

The use of automobiles became popular in the late 19th century after invention of internal combustion (IC) engines. The use of fossil fuels in IC engines proved to be more efficient and practical than steam engines available at that time. In the 1950s, however, occurrence of photochemical smog in major cities drew attention of authorities to air pollution issues. About a decade later, the US state of California approved the first legislation imposing emission limits for gasoline engines. The first European legislation was established in the following decade. Since then, new regulations were created and emission limits drastically reduced, forcing the automotive industry to continuously develop novel technologies to improve emissions control systems.

An automotive exhaust system is constituted generally of the following components: exhaust manifold, exhaust pipe, catalytic converter, particles filter, muffler and resonator, and tailpipe. This system is assisted by other secondary components like heat shields, sensors and clamps, brackets and hangers, which are also relevant for the final performance. Among these, the most relevant parts in the scope of this work are the exhaust pipe and the catalytic converter. The exhaust pipe conducts the hot gases collected from the engine by the manifold to the catalytic converter. These parts are usually made of aluminized steel, stainless steel, or zinc-plated heavy-gauge steel to avoid corrosion and oxidation [Erja2000], whereas some high performance vehicles are equipped with titanium exhaust pipes, due to its excellent performance and low weight

[HiTh2012]. The catalytic converter is responsible for conversion of harmful combustion products into more environmentally friendly gases. It is most commonly composed of catalytic-active elements like platinum, palladium and/or rhodium supported by a ceramic honeycomb-like monolith or ceramic pellets, with an $Al_2O_3$ washcoat to increase surface area. The whole catalytic converter is then packed and sealed in a steel shell [Erja2000]. The required use of noble metals (1 to 5 g [Reif2015]) makes catalytic converters the most expensive parts in an automotive exhaust system.

### 2.4.1 *Automotive emissions and European legislations*

A complete combustion of fossil fuels results in the emission of nitrogen ($N_2$), water ($H_2O$) and carbon dioxide ($CO_2$), among which only $CO_2$ is considered environmentally harmful, as it is classified as a greenhouse gas, contributing to global warming. However, the achievement of adequate conditions for complete combustion of fuels would demand the use of engine designs with reduced overall performance and increased fuel consumption. After the development of catalytic converters, engineers were able to tune engine designs for performance and lower fuel consumption, sharing the emissions control function with the exhaust systems. Hence, combustion of fuels under real conditions results in the emission of undesired compounds as well. Unburnt hydrocarbons (HC), which include also volatile organic compounds (VOC), and carbon monoxide (CO) originate from low air/fuel ratios ($\lambda_{a/f} < 1$)[1], which cause an insufficient availability of oxygen to complete fuel combustion. Unburnt HC result as well from crevices in the combustion chamber, which trap the fuel mixture, preventing complete burning, or from adsorption and desorption phenomena at the oil film in the combustion chamber. Inadequate air/fuel ratios cause additionally the formation of particulate matter (PM) or soot in Diesel engines and in Otto engines with direct injection (injection of fuel directly in the combustion chamber), due to localized oxygen deficiencies and consequently thermal cracking of hydrocarbons at high temperatures (above 1200 °C [Reif2014]). Furthermore, formation of oxides of nitrogen (NOx, with X = 1, 2) is related to excessively high combustion temperatures (above 1093 °C [HiTh2012]), which lead to the combination of nitrogen and oxygen atoms.

The great number of studies pointing out the contribution of greenhouse gases, like $CO_2$ and methane, to global warming increased the concern about such emissions. Nevertheless, the most dangerous phenomena, to which automotive emissions

---

[1] The air/fuel equivalent ratio $\lambda_{a/f}$, also known as excess air factor, is the value resulting from the division of the real air/fuel ratio by the stoichiometric air/fuel ratio (approx. 14.7 kg of air to 1 kg of gasoline [Reif2015], or 14.5 kg air for each kg of diesel [Reif2014]).

contribute, are photochemical smog and acid rain. Photochemical smog is formed by reaction of NOx with VOCs, triggered by sunlight, resulting in particulate matter and ground-level ozone. Acid rain occurs by reaction of NOx and SO₂ with water in the atmosphere, to form nitric and sulfuric acids, respectively. Fig. 2.4.1 presents the average composition in vol% of combustion gases emitted by Otto engines [Reif2015] running with a stoichiometric ratio ($\lambda_{a/f}$ = 1). Although the values of pollutants emission do not seem to be expressive, the high number of vehicles in activity makes them significant – only in Germany, 54.6 million motorized vehicles were registered by January 2016 [ZFZR2016]. Hence, the environmental impact of automotive emissions has been intensively studied in the last decades. According to the European Environment Agency (EEA), road transportation is responsible for 39% of NOx emissions, 12% of the VOC, 22% of the CO, and 12.5% of the particulate matter emissions in Europe [EEA 2015].

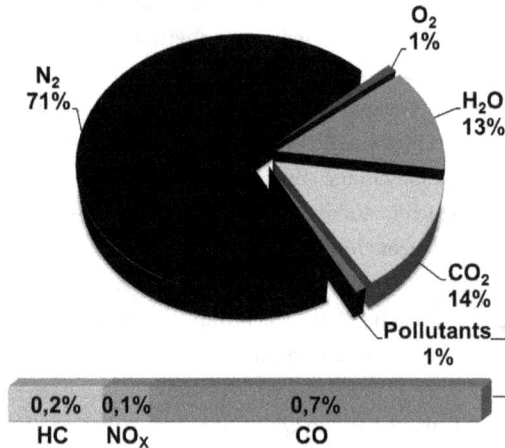

Fig. 2.4.1: *Average gaseous emissions of an Otto engine (gasoline) running with stoichiometric air/fuel ratio, in vol%. Based on [Reif2015].*

Aside from CO₂, which is only considered environmentally harmful, the pollutants in automotive exhaust gases are also damaging to human health [LlCa2001]. Additionally to odor, unburnt HC may cause irritation to eyes and nose and are considered carcinogenic. Carbon monoxide is a poisonous gas, which adheres to hemoglobin, preventing adequate oxygen transport through the circulatory system. NOx may react with water to form acids responsible for several lung disorders. It may also cause irritation of eyes and nose and may affect the nervous system. Moreover, particulate matter is damaging to the lungs, causing breathing issues, and may act as carrier to unburnt hydrocarbons into the alveoli.

As a consequence of environmental and health hazards, imposition of more severe emissions legislations forces the automotive industry toward improvements in emissions control. In Europe, the first legislation was the Directive 70/220/EEC [Euro1970] from 1970, which imposed emission limits only for Otto engines. In 1992, the Euro 1 norm [Euro1991] was implemented and in the last 23 years, this legislation has been frequently updated, reducing emission limits. The Euro norms distinguish emission limits based on type of internal combustion engine and type of vehicle [Euro2002]. According to the type of IC engine, vehicles are divided in positive ignition (PI) for engines requiring a spark to ignite the fuel (Otto engines, which operate with gasoline, ethanol, natural gas, LPG, etc.) or compression ignition (CI) for vehicles with engines, which do not require a spark to ignite the fuel (Diesel engines). Based on size and function, the European legislation divides automotive vehicles with at least four wheels into three basic categories, among which category $M_v$ stands for passenger vehicles and $N_v$ for cargo vehicles.

The evolution of emission limits for passenger vehicles through the years is presented in Table 2.4.1.

*Table 2.4.1: Emission limits for passenger vehicles (category $M_v$) according to the Euro norms for positive ignition (PI) and compression ignition (CI) engines.*

| Year | Legislation | | CO | HC | NOx | NOx + HC | PM |
|------|-------------|-----|------|-----|-----|----------|-----|
| | | | [mg km$^{-1}$] | | | | |
| 1992 | Euro 1[a] | PI | 2720 | - | - | 970 | - |
| | | CI | 2720 | - | - | 970 | 140 |
| 1996 | Euro 2[b] | PI | 2200 | - | - | 500 | - |
| | | CI | 1000 | - | - | 700/900* | 80/100* |
| 2000 | Euro 3[c] | PI | 2300 | 200 | 150 | - | - |
| | | CI | 640 | - | 500 | 560 | 50 |
| 2005 | Euro 4[c] | PI | 1000 | 100 | 80 | - | - |
| | | CI | 500 | - | 250 | 300 | 25 |
| 2009 | Euro 5[d] | PI | 1000 | 100 | 60 | - | 5* |
| | | CI | 500 | - | 180 | 230 | 5 |
| 2014 | Euro 6[d] | PI | 1000 | 100 | 60 | - | 5* |
| | | CI | 500 | - | 80 | 170 | 5 |

*Vehicles with direct injection                Sources: [a][Euro1991], [b][Euro1996], [c][Euro1998], [d][Euro2008]

In the norms Euro 1 and Euro 2 [Euro1996], more attention was given to CO and PM emissions, whereas emissions of NOx and HC were not individually taken into account. Limits for CO emissions were considerably updated since then, with a reduction of 62% for PI and 82% for CI engines from Euro 1 to Euro 4 [Euro1998], with no updates in Euro 5 and Euro 6 [Euro2008].

The pollutant with the greatest reduction of emission limits was the particulate matter, for which the limit for vehicles in category $M_v$ and CI engines decreased from 140 mg km$^{-1}$ in Euro 1 to only 5 mg km$^{-1}$ since Euro 5, which represents a reduction of about 96%. The increasing use of direct injection systems, which generate considerably more PM than engines with conventional injection systems, forced legislation to determine PM emission limits also for PI engines, starting in Euro 5. The maximum allowed emissions are the same as those for CI engines.

Limits for HC emissions (which account not only for unburnt hydrocarbons in exhaust gases but also evaporated fuel from the fuel system, for example) started to be individually determined since Euro 3, although only for PI engines. After a 50% reduction from Euro 3 to Euro 4, the limit for passenger vehicles was kept constant at 100 mg km$^{-1}$.

Like unburnt hydrocarbons, NOx emissions have been mentioned individually since Euro 3 and have gained relevance in legislation ever since. The NOx emission limit for passenger vehicles with PI engines was reduced by 60% in the period of 9 years from Euro 3 (150 mg km$^{-1}$) to Euro 5 (60 mg km$^{-1}$), whereas no updates were made in Euro 6 for this type of engine. For CI engines, the limit was reduced from 500 mg km$^{-1}$ in Euro 3 to only 80 mg km$^{-1}$ in Euro 6, a reduction of 84% in the last 15 years. Table 2.4.2 presents a comparative of NOx emission limits determined by Euro 5 and Euro 6 for passenger and light/medium cargo vehicles with CI engines. As one can observe, the emission limit for NOx was reduced in almost 55% from Euro 5 to Euro 6, which represents a great technological challenge for the automotive industry.

Table 2.4.2: Comparison of emission limits for categories $M_v$, $N_{v1}$ and $N_{v2}$ according to Euro 5 and 6 for compression ignition (CI) engines.

| Category | Class | Euro 5 [mg km$^{-1}$] | | Euro 6 [mg km$^{-1}$] | |
|---|---|---|---|---|---|
| | | NOx | NOx + HC | NOx | NOx + HC |
| $M_v$ | - | 180 | 230 | 80 | 170 |
| $N_{v1}$ | I | 180 | 230 | 80 | 170 |
| | II | 235 | 295 | 105 | 195 |
| | III | 280 | 350 | 125 | 215 |
| $N_{v2}$ | - | 280 | 350 | 125 | 215 |

Source: [Euro2008]

### 2.4.2    Methods for emissions control

In order to attend the requirements of legislations, the automotive industry is continuously searching for new technologies to reduce emissions. As mentioned earlier, the control of emissions may be managed by changing design or operation parameters of engines, in order to prevent emissions, or it may be addressed after combustion by means of the catalytic converter and other countermeasures. Among all pollutants generated by combustion of fossil fuels (CO, $CO_2$, HC, NOx and PM), only $CO_2$ emissions cannot be reduced by a more effective exhaust system, as they depend on fuel consumption and engine efficiency. In fact, an efficient exhaust system increases $CO_2$ emissions by converting harmful CO and HC into less harmful $CO_2$.

As mentioned in the previous section, formation of NOx originates from high combustion temperatures (above 1093 °C [HiTh2012]), caused mainly by high flame speed in the combustion chamber. At such high temperatures, oxygen and nitrogen combine, forming the oxides. In contrast to other pollutants, NOx emissions increase with air/fuel ratio close to the stoichiometric value ($\lambda_{a/f} \geq 1$), whereas the highest amount of NOx is generated with ratios of approximately 16:1. Furthermore, the majority of the theoretically optimal conditions for complete combustion and, consequently, for reduction of emission of the other pollutants, favor the formation of NOx. Thus, control of NOx emissions is difficult and has demanded a great effort from the automotive industry.

*Catalytic converter*

In the 1980s, the first catalytic converters capable of converting all three types of pollutants (HC, CO, NOx) were introduced, the so-called three-way catalytic converter. Due to the different nature of the pollutants, two types of catalysts are necessary: a reduction catalyst (e.g. rhodium) for conversion of NOx gases into $N_2$ and $O_2$, and an oxidation catalyst (e.g. platinum or palladium) for conversion of unburnt HC into $CO_2$ and $H_2O$, and CO into $CO_2$ [ShMc2000].

Oxygen for the oxidation reactions is provided by the exhaust gas itself, if enough is available (high $\lambda_{a/f}$) or by reduction of NOx, if low $\lambda_{a/f}$ values are applied. The necessity of an eventual injection of additional air in the catalytic converter is continuously verified by the lambda sensor [Erja2000].

In contrast to the relatively simple oxidation of HC and CO, the reduction of NOx demands additional systems, especially if the engine operates with lean mixtures ($\lambda_{a/f} > 1$) [ShMc2000]. Thus, two systems were developed to assist the NOx conversion: the NOx storage catalyst (NSC) and the selective catalytic reduction (SCR) systems [Reif2014]. The NSC system reversibly stores $NO_2$ in carbonates or oxides (mostly of potassium, calcium, strontium, barium, zirconium or lanthanum), preventing NOx emissions during lean-burn operation. However, the storing process occurs satisfactorily only at temperatures between 250 and 450 °C and the NSC system must be regenerated to avoid saturation. In contrast to NSC, the SCR system is a continuous process and does not require alterations in engine operation, thus maintaining constant performance. SCR is a well-established industrial denitrification process, which applies reduction agents, such as urea or ammonium carbamate, which preferentially reduce $NO_x$, even in the presence of $O_2$.

The invention of three-way catalytic converters, NSC and SCR systems were breakthroughs in emissions control technologies. Under stabilized operating conditions, conversion rates are close to 100%. However, they still have limitations related especially to the cold start behavior of exhaust systems [WFPG2003, YuKi2013]. Since catalytic converters require a certain amount of heat to work properly, up to 90% of the total emissions in a test cycle occur during the warm-up phase [Reif2015]. The temperature required to achieve 50% of conversion is called light-off temperature and the period of time necessary to warm the catalysts up to this value is called light-off phase. The light-off temperature of currently used three-way catalytic converters lies around 300 °C, whereas full conversion occurs usually between 400 and 800 °C.

During the first seconds of operation, hot gases leaving the combustion chambers lose heat to the components of the exhaust system and are delivered to the catalytic converter with insufficient energy to reach the light-off temperature. In order to overcome this problem, several strategies have been tested. Once again, modifications in engine parameters or in the exhaust system itself may be carried out. Changes in engine operation aim to increase exhaust gas temperature and include retardation of ignition and increase in exhaust gas mass flow. However, these changes are associated, in most of the times, with efficiency loss and increased fuel consumption, which make changes in the exhaust system more interesting. Hence, manufactures of catalyst supports are continuously searching for new materials with increased thermal conductivity, to accelerate warm-up.

Furthermore, heat loss caused by heat exchange between exhaust system parts and environment may be reduced by means of thermal insulation on the outside of the parts. Indeed, double-shell systems, thermal barrier coatings and other types of insulation are able to reduce the overall heat loss. A heat exchange between exhaust gas and exhaust system parts, in contrast, cannot be avoided by these methods.

Some manufacturers tried to bring the catalytic converter closer to the manifold, which would reduce heat loss until the catalyst is reached. However, this strategy introduces problems due to high overall temperatures, which reduce lifetime of the converter. Moreover, some vehicles are equipped with additional catalytic converters with higher content of active metals and/or higher channels density that are assembled either into the exhaust manifold or next to it. This approach requires the use of additional amounts of precious metals, considerably increasing the costs of the exhaust system.

Another investigated approach was to use additional electric heating directly at the catalytic converter. In this case, the issues were related to an insufficient capacity of the 12 V electrical systems and to the unpractical waiting time, necessary to warm-up the catalytic converter before engine start. Additional fuel burners at the entrance of the catalytic converter were also experimented, but difficulties with proper mixing of gases with different temperatures are a great drawback.

Another strategy to reduce NOx emissions is the mixing of combustion products with fresh air/fuel mixture in the combustion chamber. This is carried out either by means of an exhaust gas re-circulation (EGR) system or by increasing the overlap period, during which both admission and discharge valves are open during the induction phase. This is beneficial not only to reduce combustion temperature, but also to finish combustion of CO, unburnt HC and soot.

In summary, despite the great number of proposed solutions, the problem of the light-off phase, especially regarding NOx emissions, persists and is likely to be addressed by a combination of different strategies rather than by a single approach.

## 3  EXPERIMENTAL PROCEDURES

The development of PDC-based coating systems is a complex process, which begins with the selection of components. Then, a suitable deposition method must be selected and the respective parameters adjusted. At last, a suitable processing to realize the polymer-to-ceramic conversion must be determined. However, these steps are closely interconnected. Indeed, depending on the selected materials, some processing methods may not be possible and others may be imperative. Thus, in order to obtain a suitable combination of materials and process parameters, each processing step must be carefully considered. Even with a thorough analysis of each component, interactions between these components are sometimes difficult to predict. Hence, the characterization of obtained coatings regarding properties relevant to application may reveal the necessity of further adjustment of composition, deposition parameters and/or pyrolysis conditions to improve performance.

In this chapter, firstly the criteria for the selection of components for coating systems, as well as their role during processing and application, are presented. Then, methods and parameters for deposition and pyrolysis of the coatings are elucidated. To conclude, the applied characterization methods and their respective parameters and objectives are presented.

### 3.1  Selection of materials

PDC processing is a versatile method for preparation of ceramic coatings for a broad variety of applications. Although a great amount of preceramic polymers with excellent properties have been developed in the last years, they do not enable deposition and pyrolysis of single layers thicker than a few microns, whereas several repetitions of the whole process are required to reach higher coating thicknesses. Use of fillers is the simplest approach to obtain thicker ceramic coatings by PDC processing. Moreover, some applications are only possible by the use of functional fillers with suitable properties. However, also the substrate is an important factor to be considered, as it influences the selection of coating materials and their processing as well as the behavior of the system during application. All materials used in this work were selected and characterized according to their role in the system.

#### 3.1.1  Substrate

Although the development of a coating system seems to be mainly focused on the coating itself, the role of the substrate must not be underestimated. In general, the objective of the application of a coating is to modify the surface of a specific structural material in order to improve its performance under given application conditions. This means that the coatings have to adapt to the properties of pre-selected substrates and little flexibility regarding this choice is given. Thus, the development of coatings begins with the evaluation of suitable substrates for the intended application – in the present case, metals used to produce exhaust pipes. However, different substrates, even when used for the same application, might have significantly different properties, which influence the processing of coatings differently. Hence, the development of PDC coatings is, in most cases, oriented to a specific substrate, whereas a change of substrate usually requires also significant changes in coating composition and/or processing.

Although other types of metal have been used in automotive exhaust systems, temperature resistant stainless steels are the most common. Major requirements for the steels are low material cost, sufficient thermal, oxidation and corrosion resistances to withstand the effects of hot exhaust gases and environmental conditions, as well as suitable conformability, weldability and mechanical properties. Among others, the ferritic stainless steel AISI 441 (material number 1.4509) and the austenitic stainless steel AISI 309 (material number 1.4828) fulfill those requirements and have been applied by the automotive industry to produce exhaust pipes and other exhaust system components. Table 3.1.1 presents the standard elemental compositions of the selected steel grades in weight percentage [DIN1999, DIN2014].

Table 3.1.1: Standard compositions (in wt%) of stainless steel grades 441 (ferritic) and 309 (austenitic).

| Steel | wt% | C | Si | Mn | P | S | Cr | Ti | Nb | Fe |
|---|---|---|---|---|---|---|---|---|---|---|
| AISI 441 | min | - | - | - | - | - | 17.5 | 0.1 | (3×C)+0.3 | Bal. |
| [DIN2014] | max | 0.03 | 1.0 | 1.0 | 0.04 | 0.015 | 18.5 | 0.6 | 1.0 | Bal. |

| Steel | wt% | C | Si | Mn | P | S | Cr | Ni | N | Fe |
|---|---|---|---|---|---|---|---|---|---|---|
| AISI 309 | min | - | 1.5 | - | - | - | 19.0 | 11.0 | - | Bal. |
| [DIN1999] | max | 0.2 | 2.5 | 2.0 | 0.045 | 0.015 | 21.0 | 13.0 | 0.11 | Bal. |

The steel grade 441 is a ferritic stainless steel dual-stabilized by titanium and niobium, with 17.5-18.5 wt% chromium. The austenitic grade 309, on the other hand, contains 19-21 wt% chromium and 11-13 wt% nickel. The higher chromium content of austenitic steel grade 309 is responsible for a higher oxidation and corrosion resistance. However, in comparison to austenitic grades, ferritic grades passivate and heal faster due to the higher diffusivity of chromium in the non-compact body-centered cubic structure [GTDM2004]. Moreover, absence of nickel contributes to lower material costs. However, the most important criterion for the selection of a substrate in for this work is the linear coefficient of thermal expansion. Table 3.1.2 presents the CTE and other properties of the investigated steels. While the steel grade 309 has a CTE up to 400 °C of $17.5 \times 10^{-6}$ K$^{-1}$ and $18.0 \times 10^{-6}$ K$^{-1}$ up to 600 °C, the steel grade 441 has only $10.5 \times 10^{-6}$ K$^{-1}$ up to 400 °C and

$11.5 \times 10^{-6}$ K$^{-1}$ up to 600 °C. Ceramics derived from Si-based precursors have low CTE (up to $4 \times 10^{-6}$ K$^{-1}$ [CMRS2010, GSGW2011]) and the CTE of TBC components rarely exceeds $13 \times 10^{-6}$ K$^{-1}$. Thus, the most suitable substrate to avoid excessive thermal stresses in the system caused by CTE mismatch is the steel grade 441, which was selected as substrate for the development of thermal barrier coatings by PDC processing.

Table 3.1.2: Physical and thermal properties of stainless steel grades 441 (ferritic) and 309 (austenitic).

| Properties | AISI 441[I] [DIN2014] | AISI 309[II] [DIN1999] | Unit |
|---|---|---|---|
| Density | 7.7 | 7.9 | g cm$^{-3}$ |
| CTE | 10.0 (20-200 °C) | 16.5 (20-200 °C) | $\times 10^{-6}$ K$^{-1}$ |
| | 10.5 (20-400 °C) | 18.0 (20-600 °C) | $\times 10^{-6}$ K$^{-1}$ |
| | 11.5 (20-600 °C)[III] | 19.5 (20-1000 °C) | $\times 10^{-6}$ K$^{-1}$ |
| Thermal conductivity (20 °C) | 25.0 | 15.0 | W m$^{-1}$ K$^{-1}$ |
| Young's modulus | 220 (20 °C) [IV] | 196 (20 °C)[V] | GPa |
| | 175 (500 °C) [IV] | 150 (600 °C)[V] | GPa |
| | 100 (1000 °C)[IV] | 120 (1000 °C)[V] | GPa |
| Heat capacity (20 °C) | 0.46 | 0.50 | J g$^{-1}$ K$^{-1}$ |
| Poisson's ratio | 0.3[VI] | | |

[I][Deut2015]; [II][Thys2011]; [III][ISSF2007]; [IV][Outo2012]; [V][Outo2016]; [VI][Acci2013]

Aside from the application in automotive exhaust systems, the steel grade 441 has been investigate as interconnect material in solid oxide fuel cells, also due to its low CTE and its lower cost compared to other stabilized steels, especially developed for this application [MGDR2002, CWAT2007, RGWD2008, JaCS2010, GrFS2014]. The steel grade 441 (Deutsche Edelstahlwerke GmbH, Germany) used in this work was cold-rolled into metal sheets with thicknesses of 0.20 or 1.5 mm and surface quality 2B (Eisen & Sanitär M. Bauer GmbH, Germany). Additionally, steel pipes with internal diameter of 53 mm and wall thickness of 1 mm were provided by Faurecia Emissions Control Technologies Germany GmbH, partner in the development of the PDC-based TBCs.

### 3.1.2 Preceramic polymers

The precursor is the coating component, which enables the PDC processing. Several Si-based preceramic polymers have been applied as precursors for a great number of applications, like bulk parts, fibers, CMCs and coatings. The most common types of precursors are siloxanes, carbosilanes and silazanes (see Fig. 2.1.1). Siloxanes will always lead to ceramics containing silicon and oxygen. Carbosilanes and silazanes, on the other hand, enable preparation of oxygen-free ceramics, like SiC, Si$_3$N$_4$ and SiCN. Indeed, despite the different molecular structures and compositions, all three types lead to ceramics containing oxygen upon pyrolysis in air. However, silazanes are usually more reactive than siloxanes, enabling a stronger adhesion to most substrates and filler particles, and a faster cross-linking after shaping process. Furthermore, they are in general less expensive and more readily available than carbosilanes. Additionally, most silazanes have remarkably high ceramic yield after pyrolysis in air (above 80%).

In this work, a silazane was applied as precursor in the PDC-based top-coat. In this layer, the silazane act as a temperature-stable binder for the filler particles. The solubility of silazanes in suitable organic solvents enables the preparation of coating suspensions, which can be deposited onto substrates by simple wet deposition techniques. For this application, the organosilazane Durazane™ 1800 (Merck KGaA, Germany), formerly known as HTT 1800, was selected as precursor – for the sake of brevity the trademark symbol is suppressed from here on. This material is a colorless liquid, which is synthesized by coammonolysis reaction of dichloromethylvinylsilane and dichloromethylsilane [FSKH2013]. Compared to other silazanes reported in technical literature, Durazane 1800 stands out due to its commercial availability, high ceramic yield and lower price. To accelerate cross-linking reactions, 3 wt% of dicumyl peroxide (DCP, Sigma-Aldrich GmbH, Germany) is added to the liquid Durazane 1800. Addition of the initiator DCP reduces the onset temperature of cross-linking reactions, also reducing evaporation of oligomers and thus increasing the ceramic yield. At temperatures above 130 °C, DCP induces radical-initiated polymerization and hydrosilylation reactions involving vinyl groups of Durazane 1800 (Fig. 3.1.1a) [Seif2016].

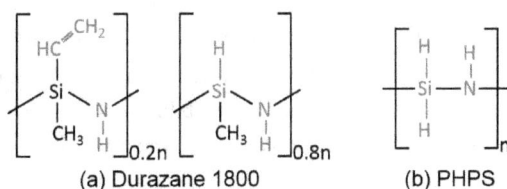

(a) Durazane 1800                    (b) PHPS

Fig. 3.1.1: *Simplified molecular structure of the silazanes (a) Durazane 1800 and (b) PHPS. The reactive groups are represented in red.*

Like traditional TBCs, some PDC coating systems, especially those developed for application at high temperatures in oxidative atmospheres, may require a bond-coat. These bond-coats are significantly thinner than the top-coat and their main task is to improve compatibility of the top-coat with the substrate. This is realized by improving adhesion and/or by increasing corrosion/oxidation resistance of the substrate. In this work, the polysilazane PHPS (perhydropolysilazane, NN120-20, Merck KGaA, Germany) was used as bond-coat material. PHPS is synthesized by ammonolysis reaction of dichlorosilane [SeWi1984], resulting in a silazane with high amount of Si–H bonds (Fig.

3.1.1b). This silazane has outstanding protective properties and its suitability as bond-coat to improve adhesion of the top-coat has been previously investigated [WGMB2011, GSGW2011, WGMF2013].

### 3.1.3 Fillers

The use of fillers usually aims at the development of thick coatings and/or preparation of coatings with functional properties, not obtained with pure PDC coatings. In the present case, the higher the coating thickness of the PDC-based TBCs, the lower are the thermal losses in the exhaust systems. Hence, the deposition of pure silazane coatings is not a suitable approach and the addition of fillers to the silazanes is required. However, only fillers with low thermal conductivity are suitable candidates for TBC systems. Moreover, the fillers are also responsible for increasing the CTE of the coating to minimize thermal stresses, especially considering that ceramics derived from silicon-based precursors have significantly lower CTEs compared to the selected steel substrate.

The most logical approach to select filler candidates for PDC-based TBCs is to consider materials, which have been applied as top-coat material in conventional TBC processing, as these materials fulfill both requirements of low thermal conductivity and high thermal expansion. As discussed in section 2.3.1, the most common top-coat material is yttria-stabilized zirconia containing 4-4.5 mol% yttria (4YSZ), especially due to the improved phase stability at very high temperatures, when compared to other varieties of stabilized zirconia. However, the application temperature intended for the PDC-based TBC system is below the transformation temperature of $ZrO_2$, and thus no phase transitions are expected during application. Notwithstanding, YSZ with 3 mol% yttria (3YSZ) possesses a lower thermal conductivity than non-stabilized zirconia owing to the presence of oxygen vacancies in the tetragonal-stabilized crystalline structure [ClPh2005]. Moreover, the thermal conductivity of stabilized zirconia is temperature-independent, while that of the monoclinic allotrope is higher at low temperatures. At last, in comparison to the 4YSZ variety, 3YSZ has higher CTE [VCTB2000, ClPh2005]. Hence, in this work, 3YSZ (H.C. Starck GmbH, Germany) with particle size $D_{90} = 0.5$ μm and $D_{50} = 0.3$ μm, determined by laser granulometry (Granulomètre 850, Cilas-Alcatel S.A., France), was used as passive filler. The main properties of 3YSZ are summarized in Table 3.1.3.

Table 3.1.3: Properties of 3YSZ [VCTB2000].

| Property | Value | Unit |
|---|---|---|
| Thermal conductivity | 2-3 | $W\ m^{-1}\ K^{-1}$ |
| Density | 5.9-6.1 | $g\ cm^{-3}$ |
| CTE (30-1000 °C) | ~11.5 | $\times 10^{-6}\ K^{-1}$ |
| $C_P$ (300-1000 °C) | 0.56-0.63 | $J\ g^{-1}\ K^{-1}$ |

The extensive shrinkage of silazane precursors may lead to severe cracking and spallation of coatings, despite the presence of passive fillers. Thus, the use of an active filler to control shrinkage may be necessary. In this case, the same criteria used to select the passive filler apply. Ideally, the active filler should generate compounds with low thermal conductivity and high CTE upon the reaction either with the precursor, with pyrolysis products or with the atmosphere. These reactions must take place only at higher temperatures to avoid premature volume increase. Moreover, a sufficient expansion of the filler particles must occur in the temperature range of pyrolysis, which is limited by the temperature/oxidation resistance of the substrate. In order to obtain suitable thermal properties, materials, which react to form $ZrO_2$, like metallic zirconium, zirconium carbide and zirconium disilicide, are promising candidates. Metallic zirconium is a very reactive material, which spontaneously reacts to form the oxide, even at room temperature. Zirconium carbide reacts with oxygen to form gaseous $CO_2$ additionally to $ZrO_2$, limiting particle volume expansion to about 41% after complete reaction. The complete oxidation of $ZrSi_2$, on the other hand, leads to formation of $ZrO_2$ and $SiO_2$. While the zirconia fraction possess the required low thermal conductivity and high CTE, the silica fraction also has a low thermal conductivity and may reduce oxygen diffusivity through the coating. The complete oxidation of $ZrSi_2$ leads to a volume expansion of the material in about 70%. This high volume expansion enables the use of low amounts of active filler to compensate shrinkage of the precursor and, consequently, allowing use of high amounts of passive filler to improve thermal properties of the system. A drawback of this material is the low CTE of the $SiO_2$ phase formed. Zirconium disilicide (HMW Hauner Metallische Werkstoffe, Germany) with particle size up to 100 μm was acquired. In order to be suitable as filler material for processing of the TBCs, the $ZrSi_2$ powder was ball-milled (MiniCer, Netzsch GmbH, Germany) with $ZrO_2$ balls with 1 mm in diameter to obtain a powder with $D_{90} = 3.0$ μm and $D_{50} = 1.5$ μm, measured by laser granulometry (Granulomètre 850, Cilas-Alcatel S.A., France).

### 3.1.4 Solvent and Dispersant

To prepare coatings, suspensions containing all components were prepared and deposited onto substrates. The solvent plays an important role, as it enables adjustment of viscosity to the selected coating deposition methods. Di-*n*-butyl ether (Acros Organics BVBA, Belgium) with purity above 99% was used as solvent in the present work. Di-*n*-butyl ether is a non-polar solvent, which does not react with the precursor. Furthermore, its relatively high boiling point and low vapor pressure ensures a low solvent loss during preparation and deposition of suspensions and a slow evaporation afterwards, improving homogeneity of the coatings. Basic properties of di-*n*-butyl ether are presented in Table 3.1.4.

*Table 3.1.4: Properties of di-n-butyl ether [Acro2012].*

| Property | Value | Unit |
|---|---|---|
| Boiling point | 141 | °C |
| Vapor pressure (20 °C) | 6.4 | mbar |
| Density | 0.768 | g cm$^{-3}$ |
| Viscosity | 0.69 | mPa s |

The dispersant DISPERBYK-2070 (BYK-Chemie GmbH, Germany) was used to prepare stable and well-dispersed suspensions. DISPERBYK-2070 is a polyacrylate co-polymer, which is soluble in di-$n$-butyl ether and does not react with Durazane 1800.

## 3.2 Compositions and preparation of coating suspensions

The composition of the coatings was varied to evaluate the effect of each component in the system. All values presented refer only to the system precursor/fillers, disregarding the fractions of solvent and dispersant. The initiator is included in the fraction of silazane.

In order to reduce the negative effects of precursor shrinkage and of the low CTE of the silazane-derived ceramic, the amount of precursor must be kept at a minimum value, sufficient to obtain good adhesion and cohesion. Hence, the amount of precursor in the systems was varied from 10 to 30 vol% in preliminary experiments.

To estimate the amount of active filler required to compensate precursor shrinkage, calculations were performed according to the active filler-controlled pyrolysis (AFCOP) methodology [GrSe1991, GrSe1992]. As discussed in section 2.1.3, this method accounts for changes in density and mass of the starting components to estimate the amount of active filler required to obtain a dense ceramic body with zero shrinkage during pyrolysis. Input parameters for the calculations are presented in Table 3.2.1.

*Table 3.2.1: Input parameters of the AFCOP calculations for the system Durazane 1800/ZrSi$_2$.*

| Parameter | Durazane 1800 | ZrSi$_2$ | Unit |
|---|---|---|---|
| Mass change in air ($\alpha$) | 0.82[I] | 1.65 | |
| Density (initial) | 1.02[II] | 4.88[III] | g cm$^{-3}$ |
| Density ratio ($\beta$) | 0.45 | 1.52 | |
| Expansion parameter ($\alpha\beta$) | 0.38 | 2.51 | |
| Linear shrinkage ($\varepsilon$) | 0.28 | -0.36 | |
| Volume fraction (zero shrinkage) | 0.70 | 0.30 | |

[I]Experimental value for Durazane 1800 with 3 wt% DCP [GSGW2011]; [II][AZEM2013]; [III][Haun2013];

The complete oxidation of ZrSi$_2$ to ZrO$_2$ and SiO$_2$ causes a mass increase of 65% – due to incorporation of oxygen – and a reduction of the overall particle density from 4.88

to about 3.21 g cm$^{-3}$. The maximum packing density of the filler particles used in the calculations was 0.5 [ToBo2008b]. Calculations resulted in a volume ratio of 70% precursor to 30% ZrSi$_2$, which is the same ratio used by Wang et al. [WGMB2011] to prepare siloxane/ZrSi$_2$ coatings.

Although the AFCOP method offers a good estimation of the necessary amount of active filler to avoid shrinkage, some aspects of the calculations must be taken into account. Firstly, these calculations assume a complete oxidation of the active filler during pyrolysis, which may not be realized under applied pyrolysis conditions. Secondly, only a binary system consisting of precursor and active filler is considered, thereby neglecting influences of passive fillers. Passive fillers reduce the overall shrinkage by reducing the volume fraction of precursor and generally cause formation of porosity in interparticle spaces, especially if high amounts of filler are used. Hence, the amount of active filler calculated by the AFCOP procedure is most likely overestimated for ternary systems with high amount of passive filler. In order to adjust the amount of active filler in the system, volume fractions of ZrSi$_2$ in the range of 5 to 15% of the total volume (considering only silazane and both fillers) were investigated. Moreover, passive filler contents varying from 45 to 80 vol% (disregarding the volumes of solvent and dispersant) were tested. Table 3.2.2 summarizes the ranges of the amounts of each component, which were investigated in this work.

Table 3.2.2: Range of the investigated amounts of precursor and fillers (disregarding the solvent and dispersant).

| Material | Investigated amounts [vol.%] |
|---|---|
| Durazane 1800 | 10-30 |
| 3YSZ | 45-80 |
| ZrSi$_2$ | 5-15 |

The respective amounts of each component were then mixed to form suspensions. However, dispersion of fillers in the presence of the precursor using high energetic methods, like ultrasound, would lead to premature cross-linking. Hence, suspensions without precursor were firstly prepared by dissolving 6 wt% of DISPERBYK-2070 (with respect to the overall mass of powder) in di-$n$-butyl ether, followed by addition of the powders and dispersion by ultrasonication and mechanical stirring with ZrO$_2$ beads for several hours. After sufficient mixing, the suspension is stable and free of agglomerates.

Final coating formulations were prepared by two methodologies (Fig. 3.2.1). In the first method, concentrated suspensions of each filler powder were prepared separately and the correct amount of each suspension was mixed. The solvent amount was then corrected by adding pure solvent. After homogenization, the silazane was added and the mixture was homogenized again by stirring to obtain the final suspension. The second method consisted in the preparation of one single suspension containing all fillers, to which the silazane was added after dispersion. The first method is more suitable for preparation of several different coating formulations in small quantities, whereas the second method is more adequate to prepare larger quantities of a single coating formulation. Hence, the first method was used in the initial phase of the work, during which different compositions were investigated. After definition of the final composition, the second method was applied to prepare further suspensions.

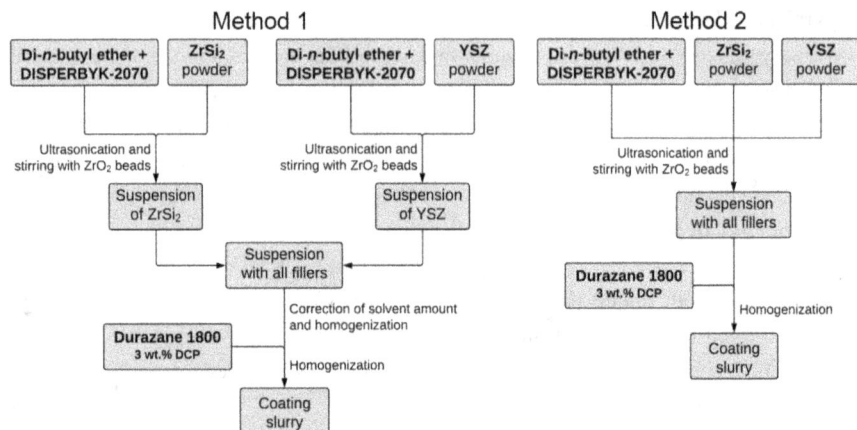

Fig. 3.2.1: Processing flowchart for the preparation of the composite coating slurries.

## 3.3   Deposition of the coatings

Different deposition techniques were used during development of the TBC system. The PHPS bond-coats were deposited onto metal sheets and pipes by dip coating technique using an automatic dip coating apparatus (RDC 15, Bungard Elektronik GmbH & Co. KG, Germany). Although the objective of the work is to deposit the TBCs onto the inner face of exhaust pipes, a protection of the exterior side against oxidation and corrosion may increase lifetime and improve appearance of the parts. Moreover, dip coating favors automation of the process and enables an easy and homogeneous deposition of thin layers. Thus, this technique is the best choice for deposition of the PHPS bond-coats. Günthner and colleagues [GKDD2009] studied the processing of PHPS coatings by dip coating technique on stainless steel substrates. According to their results, deposition of a 20 wt% PHPS solution in di-$n$-butyl ether by dip coating with hoisting speeds of 1.67-8.33 mm s$^{-1}$ yields coatings with thickness in the range of 0.7-1.2 μm. Screening tests conducted with PHPS coatings on steel 441 have shown that layers deposited with hoisting speed above 5 mm s$^{-1}$ undergo severe spallation upon pyrolysis. Hence, layers were deposited by immersing substrates in the PHPS solution and withdrawing with a speed of 5 mm s⁻¹. Drying was carried out in air at 110 °C for about 30 min.

Two different techniques were used to deposit top-coats, depending on the stage of development. In the initial phase, doctor blade technique was employed using a manual applicator frame (model 2030, BYK-Gardner GmbH, Germany) with gaps of 30, 60, 90 and 120 µm, and width of 60 mm. This method was used to deposit coatings with the different compositions onto flat steel sheets due to its simplicity, velocity and low suspension consumption.

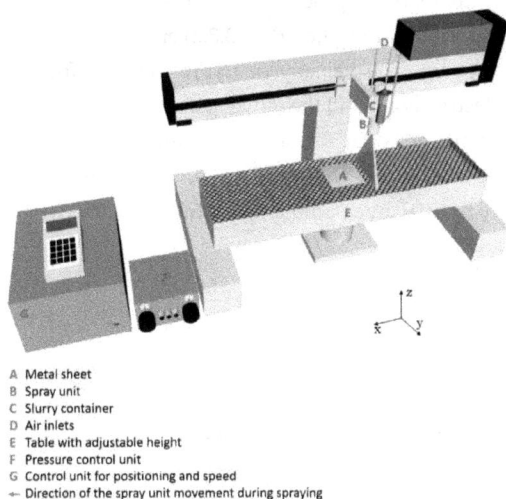

A  Metal sheet
B  Spray unit
C  Slurry container
D  Air inlets
E  Table with adjustable height
F  Pressure control unit
G  Control unit for positioning and speed
←  Direction of the spray unit movement during spraying

*Fig. 3.3.1: Schematic representation of the semi-automatic spray equipment used for the deposition of the PDC-based TBCs onto metal sheets.*

After determination of the best composition, viscosity of the coating suspensions was adjusted experimentally to spray deposition by varying the solvent amount. Deposition by spraying was carried-out using a home-made, semi-automatic spray apparatus (Fig. 3.3.1). The equipment consists of an electronic control unit, a pressure control unit, a structure to which the spraying unit (model 780S, Nordson Deutschland GmbH, Germany) with manual control of aperture is attached, and a free-standing substrate support (table), of which the position is manually adjustable. The electronic control unit enables setting of velocity (nozzle's movement) and sample's position in $x$ coordinate. The pressure control unit enables adjustment of the pressure in suspension container and of the carrier gas, as well as a manual operation of the valves. As the sample's position is set, the equipment automatically starts and stops spraying at the programmed positions. The deposition procedure is carried out with the sample on the horizontal position and the suspension is accelerated in the form of an ellipse perpendicularly toward substrate. Hence, the spray width is determined by the distance between sample and nozzle and the type of the nozzle. Deposition took place in air in a lab exhaust hood.

The deposition of coatings onto the inner face of pipes was realized using a spray gun (Perfekt 4, Krautzberger GmbH, Germany) equipped with an 1.0 m long spray lance (Krautzberger GmbH, Germany), which enables radial spraying, and a pressurized suspension container (Krautzberger GmbH, Germany). To enable a controlled deposition and better reproducibility of the process, a structure to mount the gun and the lance was developed in-house. This construction (Fig. 3.3.2) enables a remote operation of the gun via a pneumatic system. The velocity of the movement (forward and backward) is controlled by an electric motor and the movement is initiated using the remote control as well. The pipe is mounted to an adjustable structure, which enables positioning of the pipe concentric with the nozzle. The structure holding the pipe is then inserted in a spray booth (model 7475, Krautzberger GmbH, Germany) provided with filters and an exhaust system.

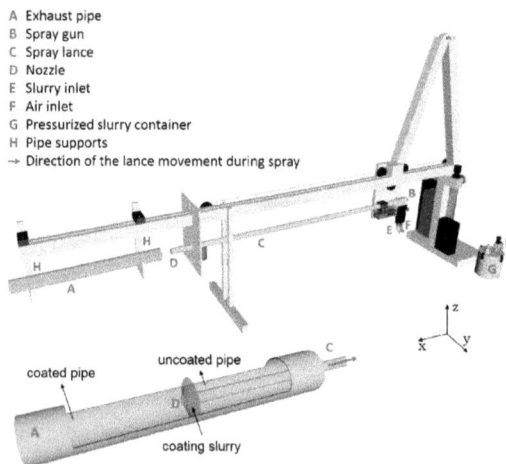

A  Exhaust pipe
B  Spray gun
C  Spray lance
D  Nozzle
E  Slurry inlet
F  Air inlet
G  Pressurized slurry container
H  Pipe supports
→  Direction of the lance movement during spray

*Fig. 3.3.2: Schematic representation of the spray equipment used for the deposition of the PDC-based TBCs onto the inner face of pipes.*

## 3.4   Thermal treatment

Conversion of the silazane-based coatings into ceramics was realized by means of thermal treatments in air. For industrial applications, a thermal treatment in air is preferred, as the use of protective atmosphere, like nitrogen or argon, increases considerably processing costs. Furthermore, as previously discussed, oxidation reactions have significantly lower activation energy and more favorable kinetic for conversion of active fillers. A disadvantage of the processing in air is the oxidation of the steel substrate, which may reduce coating adhesion and increase stresses in the coatings due to growth of an oxide scale underneath the coatings (thermally grown oxide layer).

Firstly, the PHPS bond-coat was thermally treated in a chamber furnace (Nabertherm N41/H, Nabertherm GmbH, Germany) to realize the polymer-to-ceramic conversion. Due to the high amount of Si–H and N–H groups (see Fig. 3.1.1b), PHPS is a highly reactive material in the presence of oxygen and humidity. Indeed, under these conditions, cross-linking reactions occur even at room temperature without the aid of a catalyst, mainly by hydrolysis and condensation reactions. Wang et al. [WGMF2013] have shown that the surface energy of PHPS coatings reaches its maximum at 400 °C due to increasing contribution of the surface energy's polar component, attributed mostly to formation of highly polar Si-OH groups. Above this temperature, the surface energy decreases slowly, due to condensation of Si-OH groups and formation of Si-O-Si bonds. The higher surface energy is associated with a better wettability of the surface, improving adhesion of top-coats. Furthermore, Günthner and coworkers [GKDD2009] showed by thermogravimetric analyses (TGA) up to 1400 °C in air that PHPS undergoes a mass increase, which ceases at about 450 °C and amounts to 18%. However, these measurements were performed with solid PHPS in the form of powder with particle size below 32 μm. Hence, the behavior of the powder during TGA and of coatings during pyrolysis may somewhat differ. In order to reach the state of high surface energy and at the same time to avoid further mass changes during pyrolysis of the top-coat, the bond-coat was pyrolyzed at 500 °C under air atmosphere with 1 h holding time at the maximum temperature and heating rate of 5 K min$^{-1}$.

After pyrolysis of the bond-coat, the top-coat was deposited and pyrolyzed. The presence of reactive groups Si–H, Si–vinyl and N–H in the molecular structure of Durazane 1800 (Fig. 3.1.1a) enables formation of a thermoset via condensation or addition reactions occurring up to 400 °C (thermal cross-linking) [CMRS2010]. A pyrolysis in the presence of oxygen and humidity also leads to hydrolysis and condensation reactions, increasing further the ceramic yield compared to a pyrolysis under protective atmosphere. As previously mentioned, the addition of DCP as initiator to vinyl-polymerization and hydrosilylation reactions reduces the onset temperature of cross-linking and reduces mass loss caused by evaporation of oligomers in the initial phase of the thermal treatment. In order to ensure the stability of the coatings during application in exhaust systems, where temperatures up to ~950 °C may occur, the coatings must be pyrolyzed above this temperature. Hence, cross-linking and pyrolysis were carried out in a single thermal treatment in chamber furnace (Nabertherm N41/H, Nabertherm

GmbH, Germany) in air up to 1000 °C with 1 h holding time at the maximum temperature. In the initial phase of development, a heating rate of 3 K min$^{-1}$ from room temperature up to 1000 °C was applied. After definition of the final composition, the pyrolysis program was modified to reduce pyrolysis time. A new program with a heating rate of 5 K min$^{-1}$ up to 700 °C and then a second heating step with 3 K min$^{-1}$ up to 1000 °C was then applied. Cooling down was realized naturally with the furnace closed.

Fig. 3.4.1 summarizes the complete procedures for the preparation of the PDC-based TBCs, including all deposition steps and thermal treatments.

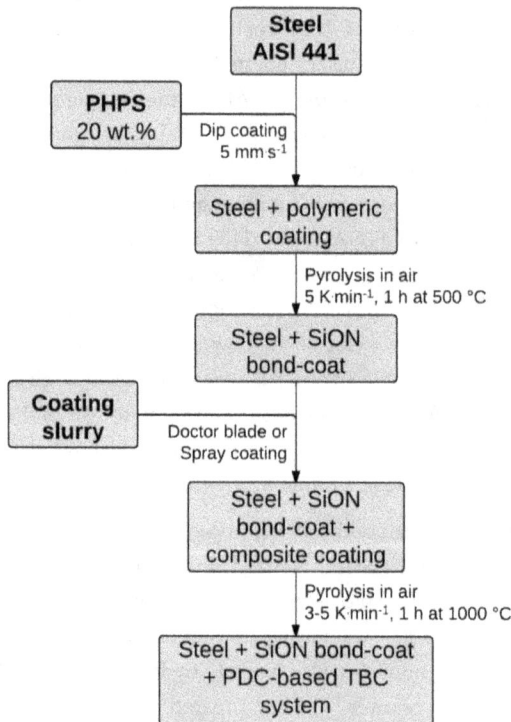

*Fig. 3.4.1: Schematic representation of the processing for the preparation of the PDC-based TBCs and structure of the PDC-based TBC system.*

## 3.5  Characterizations

### 3.5.1  *Behavior during thermal treatments*

To investigate the behavior of the silazane Durazane 1800, of the fillers 3YSZ and ZrSi$_2$, and of the final coating system during the respective thermal treatments, thermogravimetric analyses (L81 A1550, Linseis Meßgeräte GmbH, Germany) were carried out. In this technique, the mass of a sample is monitored during the defined temperature program in a selected atmosphere [HeCa1989]. Hence, the ceramic yield of precursor and the oxidation of the active filler may be analyzed to estimate the composition and the coefficient of thermal expansion of the coatings after pyrolysis. The analyses were performed in synthetic air up to 1400 °C with heating rate of 3 K min$^{-1}$ for pure components and up to 1000 °C for the final coating system. The behavior of ZrSi$_2$ upon prolonged exposure to high temperatures was also investigated by TGA in synthetic air with heating rate of 5 K min$^{-1}$ up to 1000 °C and holding time of 10 h.

A comparison of the TGA curves of pure components with that of the coating system enables an evaluation of synergistic effects between components, which may cause a change in thermal behavior of the system. In order to enable a direct comparison of results, a suspension was prepared without dispersant, maintaining the proportion of the remaining components. The solvent was removed under reduced pressure at room temperature and the obtained powder was milled with pestle and mortar.

### 3.5.2 Thermal expansion

Dilatometry technique was used to evaluate the linear CTE of steel grade 441 up to 1000 °C. Samples with 15 mm (length) x 7 mm (width) x 1.5 mm (thickness) were analyzed using a push rod dilatometer (402E, Netzsch Gerätebau GmbH, Germany) in air up to 1000 °C with a heating rate of 10 K min$^{-1}$. The linear CTE is calculated according to Eq. 3.5.1 [Cver2002], as follows:

$$\alpha^l = \frac{(l_f - l_0)/l_0}{(T_f - T_0)} = \frac{\Delta l/l_0}{\Delta T} \qquad \text{(Eq. 3.5.1)}$$

where $\alpha^l$ is the linear CTE, $l$ is the sample's length and $T$ is the temperature. The indexes 0 and $f$ indicate the initial and final values, respectively. The measurements were repeated three times with the same sample and two times more with new samples to verify possible behavior changes caused by exposure to high temperatures and oxidation.

This technique was also used to evaluate the overall dimensional changes of the coating material. In this case, samples were prepared firstly by drying the coating suspension at reduced pressure. The resulting powder was then milled in a vibratory cup mill (pulverisette 9, Fritsch GmbH, Germany) and warm-pressed with 200 MPa uniaxial pressure at 95 °C for 20 min to ensure a good cohesion an compaction of the samples, followed by another step at 160 °C for 2 h with 5 MPa uniaxial pressure to cross-link the silazane and stabilize the shape. Cylindrical specimens with 20 mm in diameter and about 2 mm in height were obtained. Dilatometry samples were prepared by cutting cuboids with dimensions of 15 mm (length) x 2 mm (width) x ~2 mm (thickness) from the cylinders. Some samples were analyzed in green state using the same temperature program defined for the pyrolysis of coatings. Other samples were firstly pyrolyzed in the chamber furnace following the same program as the coatings and only afterwards analyzed by dilatometry. In this case, the analysis program consisted of heating samples up to 1000 °C with 5 K min$^{-1}$ with holding time of 15 h at 1000 °C. Both set of samples were additionally subjected to the same program used for measurement of the linear CTE of steel samples.

The dilatometry technique measures changes in length of a sample during a given thermal treatment. In contrast to steel, the coating material undergoes not only thermal expansion/retraction but also dimensional changes caused by shrinkage of the precursor and expansion of the active filler. Hence, dimensional variations occurring during the heat-up phase of the pyrolysis program result from the combination of these three contributions. It is important to mention, however, that the CTE of the coating material changes during pyrolysis due to composition changes. Additionally, the oxidation rate of the active filler in the monolithic samples may differ from that of the particles in the coatings, due to significantly different surface areas and diffusion paths. Thus, the obtained values are only approximations to the real dimensional changes occurring during pyrolysis of the coatings.

The analysis of previously pyrolyzed specimen enables an evaluation of further expansion of the active filler after pyrolysis. This information is obtained during the plateau phase of the program. At constant temperature, no further thermal expansion occurs and a further shrinkage of the previously pyrolyzed SiCNO ceramic is unlikely. Therefore, eventual dimensional changes result from further expansion of the active filler. Moreover, within the 15 h at 1000 °C, the oxidation of the active filler should reach its completion and thus, any dimensional changes occurring afterwards are result of thermal expansion/retraction. The final linear CTE can then be obtained by subjecting these samples to the same dilatometry program used to characterize the CTE of steel samples.

### 3.5.3   Thermal conductivity

The most important property of the developed coating systems is thermal conductivity. Laser flash is the most frequently used method for thermal conductivity measurements of TBCs. However, preparation of the required free-standing samples (without substrate) of thin coatings is difficult and not well reproducible. Furthermore, laser flash technique is an indirect measurement of thermal conductivity, whereas thermal diffusivity is actually measured. The thermal diffusivity can then be converted into thermal conductivity using density and heat capacity of the samples, which may be inaccurate in the case of heterogeneous systems. Additionally, complex composites are difficult to characterize by laser flash due to the variation of the heat flow within the material [AlEK2008]. Thus, alternative methods to determine thermal conductivity of coatings have been suggested including photoacoustic and 3ω methods [TaWX1999]. The 3ω technique was applied in this work.

The 3ω method has its origins on the hot-wire technique and was presented by Cahill and Pohl [CaPo1987] as a method to determine the thermal conductivity of glasses without the negative effects of blackbody infrared irradiation. Since then, the technique has been extensively studied and applied [Cahi1990, TaWX1999, MHLG2002, WaSe2009, ASPS2010]. For measurements, a thin gold stripe is sputtered onto the coating's surface using a mask, as schematically shown in Fig. 3.5.1a. This stripe works as heater and thermometer. At the terminals of this stripe denoted "C" in Fig. 3.5.1a, an alternated current with angular frequency ω is applied. Due to the stripe's electrical resistance, heat is generated and thermal energy is transmitted from the stripe into the sample.

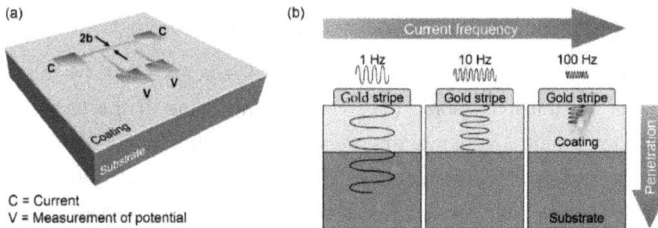

Fig. 3.5.1: Schematic representation of the 3ω method for the measurement of thermal conductivity of coatings: (a) preparation of the samples; (b) influence of the current frequency on the penetration depth of the heat waves.

The mathematical treatment to derive the thermal conductivity was described by Cahill [Cahi1990]. It starts with the solution for the amplitude of temperature oscillations at a distance $r$ from a line source of heat, given by Carslaw and Jaeger [CaJa2008]:

$$\Delta T = \frac{P}{l\pi\lambda}K_0(qr) \qquad \text{(Eq. 3.5.2)}$$

where $P/l$ is the power per unit length generated by an oscillating current with angular frequency ω passing through the metal stripe, $\lambda$ is the thermal conductivity, and $K_0$ is the modified Bessel function with order zero. The factor $1/q$ describes the penetration depth of the heat wave during one cycle of the AC power, as follows:

$$\frac{1}{q} = \sqrt{\frac{D_t}{i2\omega}} \qquad \text{(Eq. 3.5.3)}$$

where $D_t$ is the thermal diffusivity, given by $\lambda/\rho C_p$, with $C_p$ being the specific heat capacity and $\rho$ the density. This equation shows that the penetration depth of the heat wave is reduced with increased frequency, as exemplified in Fig. 3.5.1b. Thus, by varying the frequency, a signal may be generated across the thickness of the coating.

Provided that $|qr| \ll 1$, Eq. 3.5.2 for temperature variation turns into Eq. 3.5.4:

$$\Delta T = \frac{P}{l\pi\lambda}\left(\frac{1}{2}\ln\frac{D_t}{r^2} - \frac{1}{2}\ln\omega - \frac{i\pi}{4} + const\right) \tag{Eq. 3.5.4}$$

which demonstrates that the temperature variation has an imaginary (out of phase) and a real (in phase) contribution. Now, to determine the temperature variation caused by a heat source with finite width, a series of mathematical manipulations of Eq. 3.5.2 using Fourier transform with respect to the x coordinate is necessary and results in Eq. 3.5.5, as follows:

$$\Delta T = -\frac{1}{2}\frac{P}{l\pi\lambda}\left(\ln\omega + \frac{ib^2}{D_t} + const.\right) \tag{Eq. 3.5.5}$$

where b is the half stripe width (see Fig. 3.5.1a). According to this equation, the real component of the temperature variation decreases linearly with an increasing logarithm of the frequency, whereas the imaginary contribution is constant.

The temperature variation of the metal stripe is measured by the voltage across the stripe at the third harmonic ($3\omega$) – hence the name of the technique. The third harmonic component is generated because a current at angular frequency $\omega$ heats the sample at $2\omega$ and causes a temperature variation ($\Delta T_{2\omega}$), which causes a change in the resistance of the metallic stripe also at $2\omega$. The resistance variation at $2\omega$ times the excitation current at $\omega$ results in a variation of the voltage across the stripe at $3\omega$. The relation of the temperature variation with the voltage at the third harmonic is given by

$$\Delta T_{2\omega} = 2\frac{dT}{dR_e}\frac{R_e}{V}V_{3\omega} \tag{Eq. 3.5.6}$$

where $V$ is the voltage across the line at frequency $\omega$, $R_e$ is the average electric resistance of the metal stripe, $V_{3\omega}$ is the voltage at the third harmonic, and $dR_e/dT$ is the slope of the calibration curve describing the variation of the stripe's electrical resistance with the temperature. Hence, considering the slope of the real contribution of the temperature variation – or of the voltage at the third harmonic – for at least two frequency values, thermal conductivity may be obtained using Eq. 3.5.7.

$$\lambda = \frac{V^3 \ln{f_2}/{f_1}}{4\pi l R_e^2 (V_{3\omega,1} - V_{3\omega,2})}\frac{dR}{dT} \tag{Eq. 3.5.7}$$

where $V_{3\omega,1}$ and $V_{3\omega,2}$ are the voltage at the third harmonic for the frequencies $f_1$ and $f_2$, respectively.

Additionally to a direct determination of thermal conductivity, the 3ω method has the advantage of simple specimen preparation. According to Taylor, Wang and Xu [TaWX1999], results obtained by 3ω method and by laser flash are in good agreement and values may be directly compared. A disadvantage of the 3ω method is the relatively low maximum measurement temperature of about 500-600 °C. However, the aim of the work is to obtain a TBC, which can reduce heat loss during the first seconds of engine operation, when the exhaust system is still cold, to improve conversion of the pollutants. As discussed in section 2.4.2, catalytic converters reach full conversion at 400-800 °C. Thus, the temperature range of the technique is sufficient for the proposed application.

The measurements were performed at room temperature and at 500 °C in a home-made equipment (Retsch Group, Lichtenberg-Juniorprofessur für Polymere Systeme, University of Bayreuth, Germany). The thermal conductivities resulted from four measurements for each temperature.

### 3.5.4  Thermal insulation

The thermal insulation of the TBCs was verified by measuring the temperature at the outer face of a coated and of an uncoated pipe, while hot air generated by a hot air gun (PHG 530-2, Robert Bosch GmbH, Germany) flew up-stream inside the pipes. The temperature was measured with a thermocouple type K and recorded by a data-logger (PCE-T390, PCE-Deutschland GmbH, Germany). The maximum gas temperature was ~550 °C and the measurements started at room temperature. The temperature was measured at the same distance from the entrance for both coated and uncoated pipes. The thermocouple was fixated using Kapton® adhesive tape. The measurements were carried out for 120 s, which is approximately the maximum duration of the light-off phase of the catalytic converter. A schematic representation of the test configuration is shown in Fig. 3.5.2.

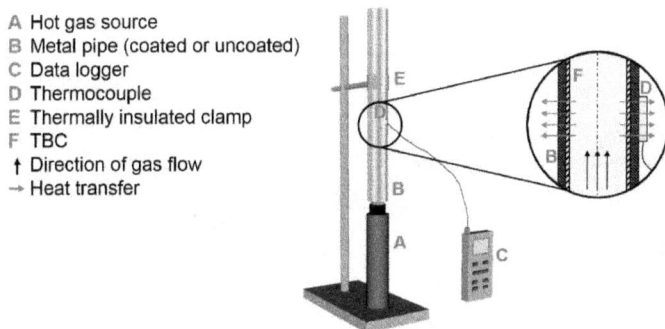

A  Hot gas source
B  Metal pipe (coated or uncoated)
C  Data logger
D  Thermocouple
E  Thermally insulated clamp
F  TBC
↑  Direction of gas flow
→  Heat transfer

Fig. 3.5.2: Schematic representation of the test of thermal insulation on pipes.

### 3.5.5  Surface roughness

The surface roughness of uncoated metal sheets was characterized by stylus profilometry. The profilometry technique characterizes roughness by scanning the surface with a diamond stylus. Displacements of the stylus caused by surface topography are detected by the equipment, which then generates a surface profile. The variations are averaged by different methods, yielding different roughness parameters. Fig. 3.5.3 presents schematically the interpretation of roughness measurements.

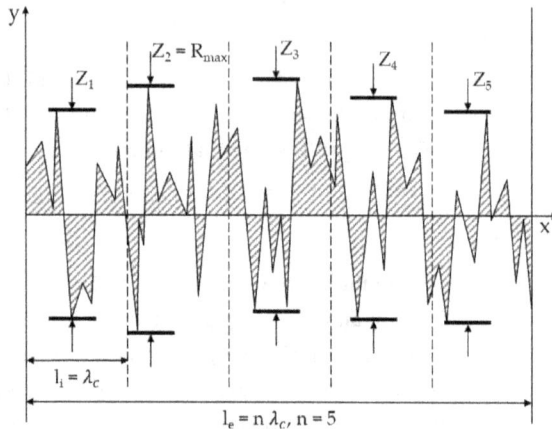

Fig. 3.5.3: Schematic representation of the profilometry measurement with the related parameters.

The most common roughness parameters are the $R_z$ and $R_a$ values, which result from the arithmetic (Eq. 3.5.8) and from the integral (Eq. 3.5.9) averaging methods, respectively. Another relevant parameter is the $R_{max}$ value, which is the maximum difference from a peak to a valley within an individual measurement length ($l_i$) unit (see Fig. 3.5.3). If $R_{max}$ exceeds the critical coating thickness, coatings may crack due to excessive coating thickness in that area. Furthermore, high roughness peaks may lead to substrate exposure, where corrosive attacks may initiate, leading to coating failure. In the present work, profilometry measurements (Garant Perthometer H2, Hoffmann Group GmbH & Co. KG) were carried out four times for each sample. The measurements were conducted with a cutoff length $\lambda_c$ of 0.8 mm, $l_i$ equal to $\lambda_c$ and evaluation length ($l_e$) of $5\lambda_c$.

$$R_z = \frac{1}{n}(Z_1 + Z_2 + \cdots + Z_n) \qquad \text{(Eq. 3.5.8)}$$

$$R_a = \frac{1}{l_e}\int_0^{l_e} |y(x)|dx \qquad \text{(Eq. 3.5.9)}$$

### 3.5.6 Coating thickness

In order to estimate the coating thickness during processing by a non-destructive technique, the magnetic induction method (ASTM B499) was applied using a coating thickness measurement gauge (Dualscope® MP40, Helmut Fischer GmbH & Co. KG, Germany). The testing probe generates a low-frequency magnetic field, of which the strength depends on the distance between probe and steel substrate. As probe and substrate are separated only by the coating, the equipment is able to determine the thickness of the coating. This technique is suitable for thickness measurements of non-magnetic coatings on ferromagnetic substrates.

### 3.5.7 Imaging techniques

Scanning electron microscopy (SEM) is one of the most important characterization techniques for development of coatings. It enables visualization of microstructure in a range down to nanometers. In microscopy, the maximum resolution of an image is limited by the wavelength of the incident radiation – the smaller the wavelength, the better is the resolution. SEM technique is based on the interaction of the sample with an electron-beam generated by a tungsten filament cathode, which has a wavelength about 100 000 times smaller than that of visible light [Schm1994]. Due to application of an electric potential difference, the electrons are accelerated toward the sample in a vacuum chamber. The impact of the electron-beam with the samples induces different types of emissions, e.g. secondary and back-scattered electrons, each of which may be detected by different detectors. By scanning the sample with the electron-beam, the topography of the surface is build up from the individual responses detected at each position or pixel. A primary requirement for SEM analyses is a sufficient electrical conductivity of the sample. To enhance the conductivity, a thin gold layer (Sputter coater 108auto, Cressington Scientific Instruments Ltd., England) or a carbon layer (Sputter coater 208carbon, Cressington Scientific Instruments Ltd., England) was sputtered on the surface of the samples. To study the microstructure of the coatings, cross-sections were prepared by cold-embedding samples in copper-containing PMMA resin followed by grinding and polishing according to typical metallographic procedures. Samples were then analyzed in a scanning electron microscope (Sigma 300 VP, Carl Zeiss Microscopy GmbH, Germany) using different detectors.

Analyses by optical microscopy were performed (Axiotech HAL 100 or Stemi SV 11, Germany, Carl Zeiss Microscopy GmbH, Germany) to evaluate results of adhesion tests and during metallographic preparation of samples for SEM analyses.

Image analysis was used to quantify porosity of the coatings as well. The graphical method employed consists in the conversion of cross-section SEM micrographs into binary images, whereas pores are represented in black and the coating material in white. The area fraction corresponding to the black regions is then quantified using the software ImageJ [ScRE2012]. By repeating these procedures on images of different regions, an estimation of porosity is possible. This method was proven to be comparable to pycnometry [TLSW2009] and offers additionally an easy and fast analysis of samples. To quantify porosity of the obtained coatings, 20 different images were evaluated and the porosity was averaged.

### 3.5.8  Density

Helium pycnometry (AccuPyc II 1340, Micromeritics Micromeritics GmbH, Germany) was used to determine the true density of fillers and of the developed coating materials. In this method, the sample is stored in a sealed chamber with a defined volume, which is filled with helium gas. Due to its small atomic size, helium is able to fill voids with dimensions in the range of angstroms. Then, a valve connecting the sample chamber with a second empty chamber opens and the system reaches a new equilibrium. By quantification of the gas displacement, by means of pressure measurements in the sample chamber, the volume of the sample may be determined with high precision. If the mass of the sample is known, density may be easily calculated. To measure density of the coating material after pyrolysis, the coating was removed from the substrate by bending coated sheets. The removed material was then milled with mortar and pestle, and analyzed.

### 3.5.9  Composition

Crystalline phases present in a material can be identified by X-ray diffraction (XRD). This method consists in the bombardment of a crystalline solid in different angles with X-rays, which are diffracted within the crystalline structure of the material. How X-rays are diffracted depends on the distances in the crystal lattice and hence, each crystalline structure of a material diffracts X-rays in a characteristic way, enabling its identification. In the present work, XRD measurements (D8 ADVANCE, Bruker AXS GmbH, Germany) were performed using CuK$_\alpha$ radiation and 2$\theta$ angles varying from 10 to 90°.

Energy dispersive X-ray spectroscopy (EDS) also relies on X-rays to identify materials. In contrast to XRD, EDS identifies elements present in a material based on the energy of the X-ray radiation emitted by the analyzed sample upon bombardment with the electron-beam in a SEM equipment. An EDS detector (EDAX Octane Super, EDAX Business Unit AMETEK GmbH, Germany) coupled to the electronic microscope was used to identify and map elements present in a selected area of the evaluated samples.

Quantitative element analyses of oxygen (Mikroanalytisches Labor Pascher, Germany) were performed to verify the presence of oxides in the $ZrSi_2$ powder.

### 3.5.10 Adhesion

The adhesion of coatings was evaluated by two standard methods. The first technique used was the cross-cut tape test (DIN EN ISO 2409 [DIN2013]), schematically represented in Fig. 3.5.4. The test consists in scratching the coating with a special tool (ZCC 2087, MTV Messtechnik oHG, Germany) to obtain a grid pattern. A piece of a suitable adhesive tape is then pressed onto the grid pattern and is removed after about five minutes in an angle close to 60°. Following, the grid pattern is visually analyzed to estimate the fraction of sample area removed with the tape. According to the amount of removed area, the coating is classified in a category ranging from Gt-0 to Gt-5, where Gt-0 is the best result, corresponding to up to 5% of removal, and Gt-5 corresponds to more than 65% of coating area removed with the tape. Typical failures observed are the removal of coating material at corners of the grid squares, at the borders or, in more severe cases, removal of the whole square. Hence, this method enables a quantitative analysis of damage rather than of adhesion itself, which is only qualitatively evaluated. Although not specified by the norm, a microscopic analysis is useful to evaluate not only the amount of removed area but also the type of failure mechanism.

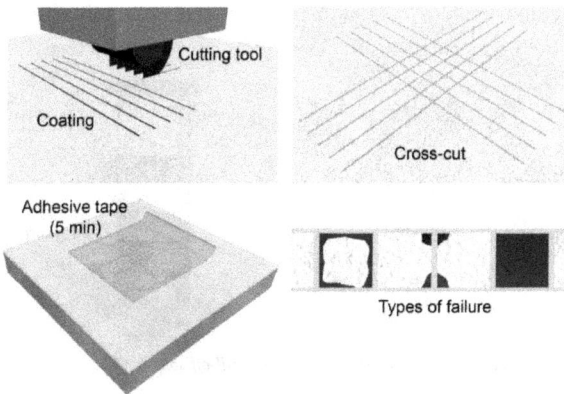

Fig. 3.5.4: Schematic representation of the cross-cut tape test (DIN EN ISO 2409).

There are two possible failure mechanisms: adhesion or cohesion failure (Fig. 3.5.5). An adhesion failure is characterized by detachment of the coating from the substrate, with the failure occurring at the interface. A cohesion failure occurs when the weakest region of the system is located within the coating and not at the interface, leading to the collapse of the coating itself. Adhesion failure is common in coatings, which adhere to substrates by weak interactions, whereas cohesion failure is the most likely to occur in porous coatings and/or layers with strong adhesion.

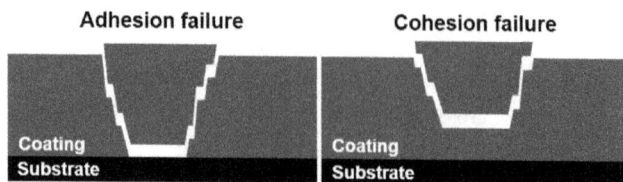

Fig. 3.5.5: Schematic representation of the failure mechanisms of coatings.

The second method applied to characterize adhesion was the pull-off test (ASTM D4541 [ASTM2009]), schematically represented in Fig. 3.5.6. To measure the adhesion strength of the coatings, an aluminium dolly is firstly glued onto the coating surface with a two-component epoxy resin (LOCTITE® EA 9466™, Henkel AG & Co. KGaA, Germany) and the sample is stored for about 60 min at 110 °C to fully harden the resin, according to the instructions of the manufacturer. The assembly is then attached to an automatic pull-off adhesion tester (PosiTest AT-A, DeFelsko Corp., USA), which perpendicularly pulls the dolly and measures the force necessary to detach it from the surface. By setting the dolly's diameter, the equipment returns the tensile adhesion of the assembly. Dollies with diameter of 10 mm and a pulling rate of 1 MPa s$^{-1}$ were applied during the tests. Thus, this method enables a quantification of the adhesion strength as well as an evaluation of failure mechanisms by posterior microscopic analyses.

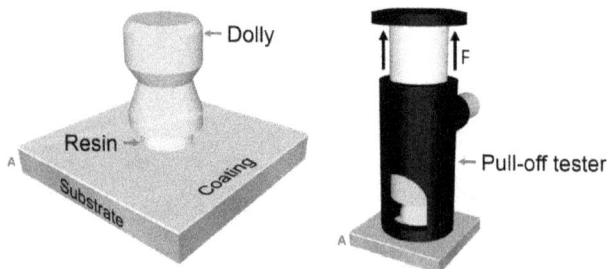

Fig. 3.5.6: Schematic representation of the pull-off adhesion test (ASTM D4541).

Adhesion failure may occur at the interfaces coating/substrate, coating/resin and resin/dolly, whereas cohesion failure is possible within the coating and within the resin. Indeed, to avoid cohesion failure within the resin layer, the resin must have a tensile strength higher than the adhesion of the coating and the layer must be free of porosity. Adhesion failure at the interfaces of the resin layer are avoided by degreasing and, whenever necessary, by roughening both the coating's and dolly's surfaces to improve resin adhesion.

### 3.5.11 Stress evolution

Different stresses occur in a PDC coating system. These stresses may be residual – i.e. generated during pyrolysis (e.g. by precursor shrinkage and expansion of active fillers) and maintained after cooling – or induced by temperature, e.g. due to CTE mismatch. If coating and substrate have significantly different thermal behaviors, the stresses may cause a deformation of substrate and/or coating, which can be evaluated by thermo-optical dilatometry technique. Considering a substrate sheet coated only on one face, if the expansion of the substrate during a thermal treatment is sufficiently higher than that of the coating, the coated face will become concave and the uncoated face convex. Analogously, if the expansion of the coating is sufficiently higher than that of the substrate, the system will curve in a way that the coated face will become convex and the uncoated face concave. However, these phenomena will only take place if the coating possess sufficient adhesion and cohesion strength to overcome the thermal stresses caused by the mismatch and if the substrate is sufficiently ductile. Hence, for coatings with the same thickness, the effects will be much more pronounced on thinner substrates.

Curvature analyses using a thermo-optical dilatometer (TOMMI Plus, Fraunhofer ISC, Germany) were performed to verify changes in the shape of coated metal sheets. The equipment consists of a temperature-resistant camera, placed in a chamber, which is heated up to the desired temperature. A software then recognizes the contours of the sample and calculates the curvature. Unfortunately, observation of the whole pyrolysis process is not possible, due to liberation of gaseous pyrolysis products, which could damage the equipment. Thus, the investigations were carried out on samples previously pyrolyzed at 700 °C for 1 h in air. The curvature investigations were conducted at temperatures up to 1000 °C with heating rate of 5 K min$^{-1}$ and holding time of 10 h.

The software of the TOMMI Plus measures the curvature $\kappa$ of the sample in pixels, whereas the curvature is defined as the inverse of the curvature radius $R_\kappa$. To enable a conversion of pixels into metric units – and thus the determination of the curvature radius – a calibration is performed with standard cylinders, which resulted in the conversion of 1 Pixel = 0.0713 ± 0.0005 mm. Using the curvature radius, the thicknesses of coating and substrate, and the Young's modulus of the substrate, the overall stress within a coating can be calculated using the Stoney's equation (Eq. 3.5.10) [Ston1909]:

$$\sigma_{St} = \frac{E_s d_s^2}{6 R_\kappa d_c}$$
(Eq. 3.5.10)

where $\sigma_{St}$ is the overall stress within the coating calculated by the Stoney's equation, $E_s$ is the Young's modulus of the substrate, and $d_s$ and $d_c$ are the thicknesses of substrate and coating, respectively. The equation initially proposed by Stoney considered only uniaxial contributions and was, therefore, only valid for parts with length much larger than the width. A modification of the original Stoney's equation was afterwards proposed and resulted in the most commonly used Stoney's equation:

$$\sigma_{St} = \frac{E_s d_s^2}{6R_\kappa(1 - v_s)d_c}$$
(Eq. 3.5.11)

where $v_s$ is the Poisson's ratio of the substrate. The Poisson's ratio was introduced to correct the Young's modulus of the substrate to account for biaxial stress conditions. It is important to mention that this equation disregards the Young's modulus of the coating and offers a good estimation of the stresses only when substrate thickness is at least 5 times larger than the coating thickness [VGMM2016].

In the present work, samples with narrow dimensions of 30 mm × 5 mm were analyzed with the TOMMI Plus, in a way that bending of the sheet took place mostly along the largest dimension. After measurement of the curvature radius, Eq. 3.5.11 was used to calculate the stresses in the coatings.

### 3.5.12 Temperature and oxidation resistance

The oxidation protection of the PHPS coating during pyrolysis in air at 500 °C for 1 h and during pyrolysis at 1000 °C for 1 h (Nabertherm N41/H, Nabertherm GmbH, Germany) was evaluated by visual inspection and comparison of the mass change of coated and uncoated samples after each thermal treatment. For these experiments, the steel was cut into large sheets with dimensions of 100 mm × 50 mm to reduce the influence of the edges on the overall mass change. The mass gain was then correlated to the surface area. The mass of coating was determined by weighing the steel sheets before and after coating deposition and drying.

Temperature and oxidation resistance of the TBC system was evaluated by subjecting coated samples with dimensions of 40 mm × 40 mm to a continuous thermal treatment (Nabertherm N41/H, Nabertherm GmbH, Germany) under air atmosphere at 900 °C with heating rate of 5 K min$^{-1}$ and holding time of 10, 50 and 100 h. Although the system was developed to endure temperatures up to 1000 °C, temperatures above 900 °C are reached only in sporadic situations with duration of a few seconds in exhaust systems of road vehicles. Indeed, average temperatures of 700-750 °C are considered by the exhaust system developers as typical for exhaust pipes located between the manifold and the catalytic converter. However, in order to investigate the phenomena taking place in a worst-case scenario, the prolonged exposure tests were performed at 900 °C. The growth of the oxide layer, the integrity of the coatings and interfaces, and changes in microstructure and composition were evaluated by SEM and EDS techniques.

### 3.5.13 Thermal shock resistance

Another fundamental property regarding durability of the TBCs is the thermal shock resistance. In exhaust systems, typical maximum temperature variations are in the range of 700-800 °C, which are caused by the contact of the hot system with cold air or with water splashes. Once again, to accelerate the effects of the thermal shock cycles, temperature variations around 1100 °C were applied. Coated samples were firstly immersed in liquid nitrogen (T = -196 °C) for 15 min, then immediately placed in a previously heated furnace at a temperature of 900 °C (LE 14/11, Nabertherm GmbH, Germany) and held for another 15 min, characterizing a full cycle. This cycle was repeated up to 30 times and the effects of the thermal shocks were evaluated by optical microscopy.

## 4 RESULTS AND DISCUSSIONS

### 4.1 Characterization of the selected materials

The comprehension of the behavior of the individual components is fundamental to explain the behavior of the coating system. By understanding how the components behave, it is also possible to identify synergistic effects, caused by interactions between the individual components of the system. Thus, different techniques were used to characterize substrate, precursors and fillers individually. Properties, such as thermal expansion, mass change, composition, surface roughness and density were evaluated.

#### 4.1.1 Properties of the substrate

The substrate has a great influence on the behavior of the coating system during pyrolysis and application. Especially surface characteristics and thermal expansion affect quality and stability of the coatings, since unsuitable roughness and excessive thermal expansion may lead to cracking and spallation of bond-coat and/or top-coat. Thus, these properties were characterized in detail.

##### Surface roughness

The surface roughness of the acquired steel AISI 441 was evaluated by profilometry technique. Four measurements were performed and the values of $R_a$, $R_z$ and $R_{max}$ where averaged and are presented in Table 4.1.1 together with the respective standard deviations ($\sigma_{SD}$). An example of the surface topography is shown in Fig. 4.1.1.

Table 4.1.1: Roughness measured by profilometry on as-received steel 441 samples with surface quality 2B.

| Parameter | Average (n = 4) [μm] | Standard deviation ($\sigma_{SD}$) |
|---|---|---|
| $R_a$ | 0.241 | 0.023 |
| $R_z$ | 1.675 | 0.193 |
| $R_{max}$ | 2.385 | 0.752 |

The acquired steel AISI 441 has a somewhat smooth surface, with low values of $R_a$ and $R_z$. However, the values of $R_{max}$ are above the critical coating thickness of PHPS coatings. In these regions with excessive roughness, the PHPS coating may become too thick and crack during conversion into ceramic. Moreover, if roughness peaks remain uncoated after PHPS deposition, oxidation may initiate at these regions and propagate underneath the coating, leading to coating failure.

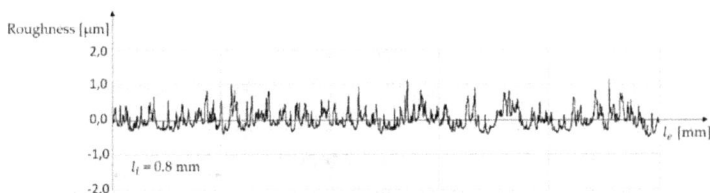

Fig. 4.1.1: Example of surface topography of the as-received steel AISI 441 sheet with surface quality 2B measured by profilometry.

Although important, quantitative roughness measurements alone give only an insight on the surface characteristics, whereas the morphology is better analyzed when SEM analyses are additionally carried out. Fig. 4.1.2 presents a SEM micrograph of the surface of the as-received steel substrate. The surface is inhomogeneous and characterized by several imperfections, among which are pores and large depressions, leading to the high $R_{max}$ values measured by stylus profilometry.

Fig. 4.1.2: SEM micrograph of the surface of an as-received steel AISI 441 sheet with surface quality 2B.

Despite this non-ideal surface characteristic, the substrates were used without further surface modification. The extent of the influence of the surface roughness on the quality of the layers is discussed in section 4.2.1.

*Thermal expansion*

Due to the lack of data from manufacturers regarding thermal expansion of the stainless steel grade 441 above 600 °C, dilatometry measurements were performed up to 1000 °C with heating rate of 10 K min$^{-1}$ in air. Fig. 4.1.3 presents CTE values calculated from dilatometry data at different temperature ranges. As one can observe, slightly higher CTE values (max. 10%) were measured in the range of 30 to 600 °C, when compared to the manufacturer's data for standard steel AISI 441 presented in Table 3.1.2. The linear CTE of the ferritic steel increases more abruptly up to 400 °C, temperature at which it reaches $11.57 \times 10^{-6}$ $K^{-1}$. Between 400 and 700 °C, the CTE increases almost linearly and reaches $12.19 \times 10^{-6}$ $K^{-1}$ for the temperature range of 30-700 °C. At the higher temperature intervals (700-1000 °C), the CTE increases again linearly, although slightly faster, and reaches the maximum value of $13.29 \times 10^{-6}$ $K^{-1}$ for the temperature range of 30-1000 °C. The same results were obtained by repeating the measurement three times with the same sample and two more with new samples. Thus, the thermal expansion behavior of the steel remains unchanged, even after exposure to 1000 °C in air.

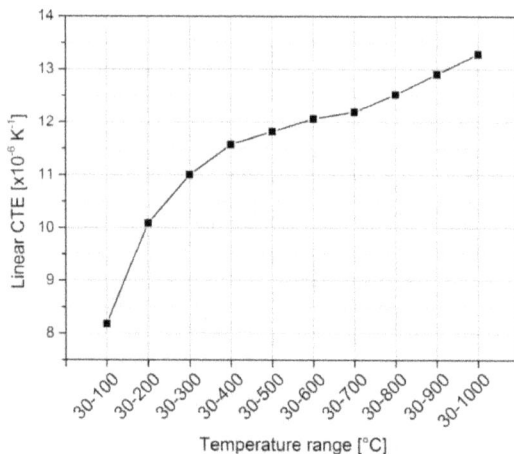

*Table 4.1.3: Linear CTE of the ferritic stainless steel AISI 441 at different temperature ranges, measured by dilatometry in air up to 1000 °C with heating rate of 10 K min⁻¹.*

### 4.1.2  Conversion behavior of the coating components

The behavior of the materials selected for preparation of the top-coat upon thermal treatment in air was characterized by TGA. Thermogravimetric curves resulting from measurements up to 1400 °C are shown in Fig. 4.1.4a.

The polyorganosilazane Durazane 1800 undergoes mass loss in three steps up to about 700 °C. No further mass change can be observed above this temperature and the total mass loss amounts to 17-18% up to 1400 °C. This result is in a good agreement with values published elsewhere [GSGW2011, MSTK2012, GPKM2014]. The transition from polymeric state to an amorphous ceramic starts with the elimination of oligomers up to 150 °C, despite the use of DCP as cross-linking initiator. Afterwards, the release of gaseous species like $NH_3$, $CH_4$, $NO_x$, $CO_2$ and $H_2O$ occurs. Motz et al. [MSTK2012] have shown that the incorporation of oxygen ceases at about 400 °C and that the product of pyrolysis up to 1000 °C is an amorphous ceramic composed of Si, N, C and O. The 3YSZ powder, as expected, did not show any significant mass change, confirming the passive filler behavior. The $ZrSi_2$, on the other hand, presented a considerable mass increase starting at about 450 °C. At 1000 °C the mass increase reaches 43% and continues to increase at higher temperatures.

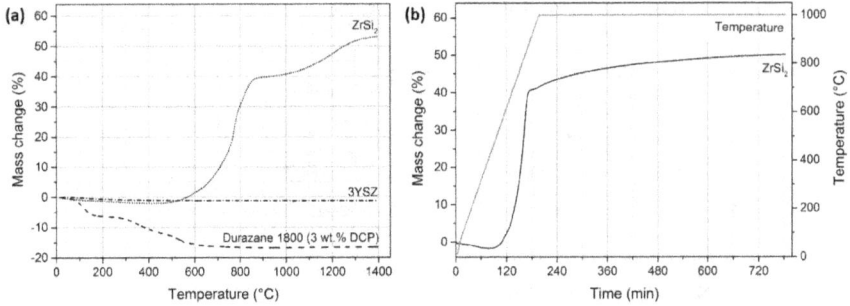

*Fig. 4.1.4: TG curves obtained by analyses in air of the top-coat components: (a) silazane and fillers, with a heating rate of 3 K min⁻¹ up to 1400 °C; (b) ZrSi₂ with a heating rate of 5 K·min⁻¹ up to 1000 °C and holding time of 10 h [BaKM2015]. Copyright © 2015 Elsevier Ltd. Adapted with permission.*

A complete conversion of $ZrSi_2$ into $ZrO_2$ and $SiO_2$ would lead to a mass increase of 65.1% and the resulting material would be composed of 28.4 vol% $ZrO_2$ and 71.6 vol% $SiO_2$, considering densities of 2.20 g cm⁻³ and 5.68 g cm⁻³ for $SiO_2$ and for $ZrO_2$, respectively. To evaluate the conversion of the active filler at the pyrolysis temperature of 1000 °C, $ZrSi_2$ was subjected to thermal treatment up to 1000 °C for 10 h. The resulting curve is presented in Fig. 4.1.4b. In order to verify the presence of previously formed oxides, elemental analysis was carried out and revealed an oxygen content of ~2.3 wt%, which means that the maximum mass increase achievable during the TGA measurements is lower than the theoretical value. As can be noticed, the mass of the filler increases continuously and reaches ~45% after 1 h at 1000 °C. After that, the mass proceeds to increase but the rate decreases significantly after about 7 h and is almost negligible after 8 h, reaching the maximum mass increase of ~50%. Geßwein et al. [GPBH2008] published a thorough study on the oxidation behavior of $ZrSi_2$. According to the authors, this material follows partially the Wagner theory of selective oxidation [Wagn1956]. This model assumes that the less noble element in the system (in this case Zr) is preferentially oxidized, while Si atoms diffuse from the forming outer oxide layer into the bulk of the material, resulting in the formation of elemental silicon, as described by Eq. 4.1.1. However, also the oxidation of silicon takes place in this first stage of mass gain, between 500 and 900 °C (Eq. 4.1.2). For this reason, the composition of the pyrolysis product is difficult to predict by thermogravimetric analysis.

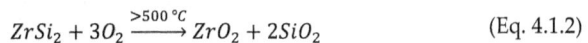

$$ZrSi_2 + O_2 \xrightarrow{>500\,°C} ZrO_2 + 2Si \qquad \text{(Eq. 4.1.1)}$$

$$ZrSi_2 + 3O_2 \xrightarrow{>500\,°C} ZrO_2 + 2SiO_2 \qquad \text{(Eq. 4.1.2)}$$

After this concurrent initial oxidation step, a further mass increase occurs above 900 °C, mainly due to oxidation of free Si to $SiO_2$. The complete pathway up to 1300 °C is described by Eq. 4.1.3. Above this temperature, the formation of $ZrSiO_4$ begins and the product of a complete oxidation of $ZrSi_2$ is a mixture of $ZrSiO_4$, amorphous and crystalline $SiO_2$ and residual $ZrO_2$ [GPBH2008].

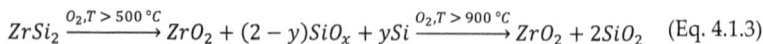

$$ZrSi_2 \xrightarrow{O_2, T > 500\,°C} ZrO_2 + (2 - y)SiO_x + ySi \xrightarrow{O_2, T > 900\,°C} ZrO_2 + 2SiO_2 \quad \text{(Eq. 4.1.3)}$$

The observed behavior during the thermogravimetric investigations is in perfect agreement with the findings of Geßwein and colleagues.

The conversion behavior of the active filler was investigated by XRD as well. $ZrSi_2$ powder as-received and after thermal treatment at 1000 °C in air for 1 h were analyzed. The respective diffractograms are presented in Fig. 4.1.5.

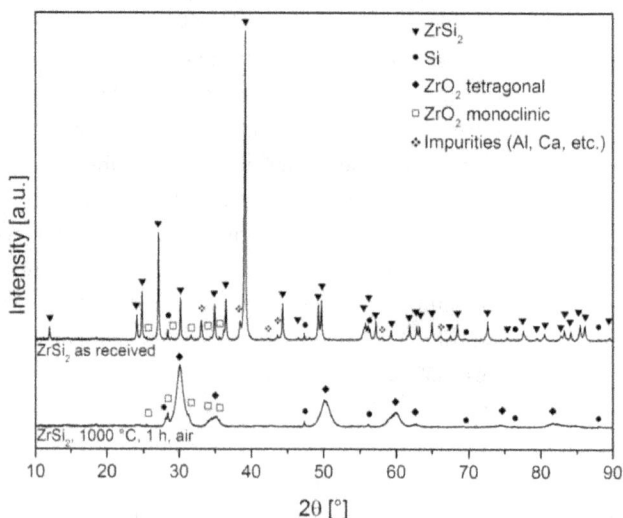

Fig. 4.1.5: XRD diffractogram of the ZrSi₂ powder as received and after treatment at 1000 °C for 1 h in air.

The diffractogram of as-received $ZrSi_2$ powder is composed of several peaks attributed to $ZrSi_2$ as well as several other peaks attributed to monoclinic zirconia, elemental Si and to impurities. As mentioned, the presence of oxides was previously verified by elemental analysis. The most typical impurities in $ZrSi_2$ are calcium, aluminium and iron [YoMI1994], which may be present in metallic form or as oxides and silicides. After thermal treatment at 1000 °C in air for 1 h, the diffractogram is composed only of peaks of monoclinic and tetragonal zirconia phases and elemental Si, which is in agreement with the oxidation mechanism proposed by Geßwein and co-workers [GPBH2008]. After thermal treatment at 1000 °C for 1 h in air, the peaks of $ZrSi_2$, even the sharp and intense one at $2\theta$ 39°, have disappeared. However, residual $ZrSi_2$ might still be present in the particle's core.

The most interesting feature of the XRD diffractogram of oxidized $ZrSi_2$ is a predominant presence of tetragonal zirconia phase, instead of the monoclinic one, as it would be expected at this temperature. Although Geßwein and colleagues obtained similar results, no explanation for this fact was provided. A hypothesis is the stabilization of the tetragonal phase by impurities in the $ZrSi_2$ powder. As previously mentioned, oxides of calcium and aluminium are stabilizers of zirconia's tetragonal phase. This also explains the absence of peaks related to impurities after thermal treatment, as they are now embedded in the crystalline structure of the tetragonal zirconia phase. Moreover, the broadness of the zirconia peaks indicates a low crystallinity of the material.

Helium pycnometry measurements were performed to determine the density of 3YSZ as received, and of $ZrSi_2$ as received and after oxidation at 1000 °C for 1 h. The density of as-received 3YSZ powder was measured to be 5.75 g cm$^{-3}$, whereas the theoretical density of tetragonal zirconia is 6.1 g cm$^{-3}$. This difference indicates the presence of some monoclinic phase (5.68 g cm$^{-3}$), which was confirmed by XRD measurements (not shown). The true density of as-received $ZrSi_2$ powder amounted to 4.84 g cm$^{-3}$, compared to 4.88 g cm$^{-3}$, which is the theoretical density of $ZrSi_2$ [GPBH2008]. The lower density in comparison to the theoretical value is related to the presence of impurities and oxides, as already observed in the elemental analysis and in the XRD measurements. The measurement of the density of $ZrSi_2$ after oxidation at 1000 °C for 1 h resulted in a density of 3.39 g cm$^{-3}$, whereas a full conversion into the reaction products $ZrO_2$ and $SiO_2$ would result in a density of 3.21 g cm$^{-3}$. As already determined by TGA and XRD measurements, the full conversion of the active filler is not achieved upon treatment at 1000 °C for 1 h in air. However, it is difficult to affirm rather only free silicon remains or if also some undetected amount of $ZrSi_2$ is still present. Considering the mass increase of 45% after 1 h at 1000 °C and the density of the oxidized powder, measured by pycnometry, it is possible to estimate the composition of oxidation product, assuming that all $ZrSi_2$ was consumed. A good approximation of these both values is obtained when $y$ in Eq. 4.1.3, which determines the amount of free silicon, is equal 0.8. For the calculations, densities of 2.20 g cm$^{-3}$ for $SiO_x$, 2.34 g cm$^{-3}$ for Si and 5.68 g cm$^{-3}$ for $ZrO_2$ were used. The estimated composition of the oxidation product of resulting from the treatment of $ZrSi_2$ at 1000 °C for 1 h in air is presented in Table 4.1.3.

Table 4.1.3: Estimated composition of the oxidation product of $ZrSi_2$ after treatment in air at 1000 °C for 1 h.

| Composition | Si | $SiO_x$ | $ZrO_2$ |
|---|---|---|---|
| Mass fractions [wt%] | 10.3 | 33.1 | 56.6 |
| Volume fractions [vol%] | 15.0 | 51.1 | 33.9 |

## 4.2   Development of the coating system

The development of a suitable coating system is challenging and requires comprehension of the behavior of each component and of the interactions occurring between them. Thus, both bond-coat and top-coat were evaluated, in order to define the final composition of the PDC-based TBC system. This system was then further characterized by different methods. The results are discussed in the following sections of this chapter.

### 4.2.1   Bond-coat

The dip coating procedure with withdrawal velocity of 5 mm s$^{-1}$ using a 20 wt% PHPS solution in di-$n$-butyl ether resulted in an average of 1.3 g m$^{-2}$ of PHPS deposited on the steel sheet – after drying at 110 °C for 1 h and considering only the coated area on both sides, neglecting the thickness of the substrate. The surface morphology of the PHPS layer was analyzed by SEM (Fig. 4.2.1).

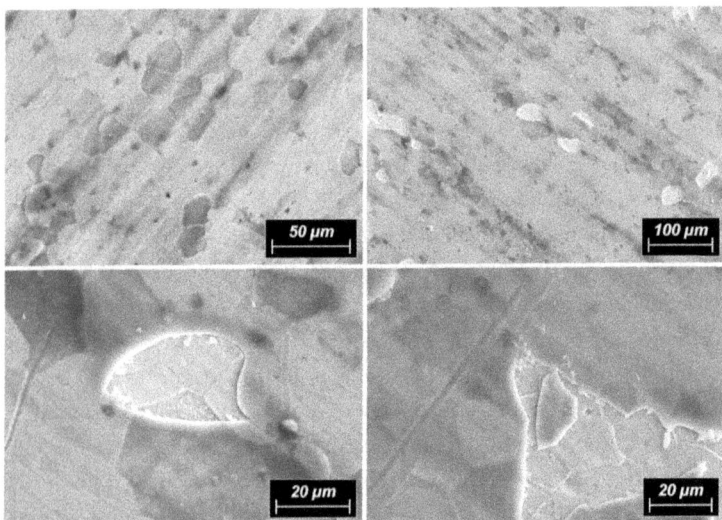

Fig. 4.2.1: SEM micrographs of a PHPS-coated steel AISI 441 sheet after pyrolysis at 500 °C for 1 h in air.

As evident in the micrographs, the PHPS-derived SiON layer was able to level the surface of the steel, covering all imperfections (see section 4.1.1). However, some localized spallation of the PHPS coating occurs. This is most likely related to the accumulation of PHPS in depressions at the surface, where the coating thickness exceeds the critical value, leading to cracking and spallation.

The effect of the bond-coat on the oxidation of the substrate was investigated by visual inspection and gravimetrically after each thermal treatment. Fig. 4.2.2a-e shows digital images of an as-received steel 441 sheet (Fig.4.2.2a), of uncoated and coated steel sheets treated at 500 °C (Fig.4.2.2b,c) and additionally at 1000 °C for 1 h (Fig.4.2.2d,e).

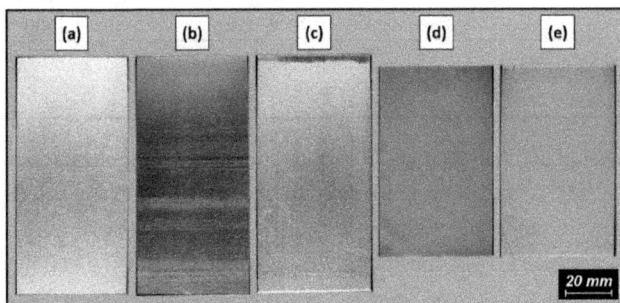

*Fig. 4.2.2: Digital images of: (a) as-received steel 441 sheet; (b) uncoated steel sheet treated at 500 °C for 1 h in air; (c) PHPS-coated steel sheet treated at 500 °C for 1 h in air; (d) uncoated steel sheet additionally treated at 1000 °C for 1 h in air; and (e) PHPS-coated steel sheet additionally treated at 1000 °C for 1 h in air.*

The oxidation of the steel 441 at 500 °C is associated with a color change from the characteristic metallic look to red/purple coloration, as evident by comparing images a and b of Fig. 4.2.2. The marks at the surface of the uncoated steel after oxidation were already present before the thermal treatment as a result of the cold-rolling process used to produce the metal sheets. When the PHPS coating is deposited onto the steel, however, the coated area maintains the metallic look (Fig. 4.2.2c), proving that, despite localized defects, the PHPS layer can protect the substrate against oxidation. The disadvantages of the dip coating technique become evident in Fig. 4.2.2c, since the upper region of the substrate – where the sheet is fixated to the dip-coater – remains uncoated and oxidizes, which is characterized by the purple coloration. Furthermore, the occurrence of deposition defects can also be observed in the lower region. In this area, the thickness of the coating exceeds the critical coating thickness and some spallation occurs, which is evidenced by the red coloration. This occurs in the lower part of the sheet because edge effects combine with the flow of solution from the upper part of the sheet. For these reasons, the upper (~5 mm from the upper edge) and lower (~15 mm from the lower edge) regions of the metal sheets were cut off after pyrolysis at 500 °C. After thermal treatment at 1000 °C (Fig. 4.2.2d,e), the oxidation protection of the PHPS coating is still evident, although the metallic look disappears. The PHPS-coated sheets acquire an opaque gray coloration while the uncoated sheet acquired an opaque dark gray/brown coloration.

The oxidation protection of the PHPS coating during pyrolysis in air at 500 °C and at 1000 °C for 1 h was additionally evaluated by weighing coated and uncoated samples before and after the thermal treatments. The mass gain was then related to the coated area. Table 4.2.1 presents the average mass gain per area of coated sample with the respective standard deviations.

Table 4.2.1: *Average mass gain caused by oxidation of uncoated and PHPS-coated steel AISI 441 sheets after treatment at 500 °C for 1 h and additionally at 1000 °C for 1 h.*

| Sample | Average mass gain (n = 2) [$\mu$g mm$^{-2}$] | |
|---|---|---|
| | 500 °C 1 h | 1000 °C 1 h |
| Uncoated steel 441 | -0.035  ($\sigma_{SD}$ = 0.007) | 1.460  ($\sigma_{SD}$ = 0.054) |
| PHPS-coated steel 441 | 0.487  ($\sigma_{SD}$ = 0.064) | 0.649  ($\sigma_{SD}$ = 0.001) |

As shown in Table 4.2.1, the substrate undergoes no significant mass change after treatment at 500 °C for 1 h, despite the color change – the negative value in Table 4.2.1 results from a difference of ~0.3 mg in the absolute weight of the samples before and after the treatment, which may be considered a measurement error. The PHPS-coated samples, on the other hand, undergo an averaged specific mass gain of about 0.487 $\mu$g mm$^{-2}$ – the area used in this case was the coated area on both sides, not the total area of the sheet. As the oxidation of the uncoated substrate leads to no significant mass gain up to 500 °C for 1 h, one can conclude that the measured mass gain is only related to the oxidation of PHPS. Despite the localized layer spallation, especially in the lower regions, the mass gain of PHPS – related to the deposited mass of coating, measured before pyrolysis – is approximately 37%, i.e. more than twice as high as the expected mass gain according to the investigations of Günthner et al. [GKDD2009], which measured a maximum mass gain of 18% on powder samples with particle size below 32 $\mu$m by TGA up to 1400 °C. Indeed, according to the same authors, the oxidation of free silicon and the complete substitution of nitrogen atoms of PHPS for oxygen should lead to a mass increase of 43%, indicating that the coatings are almost fully oxidized after pyrolysis at 500 °C for 1 h. It is well known that PHPS forms a SiO$_2$ passivating layer upon oxidation, which reduces the diffusion rate of oxygen into the core of the material. This explains why the mass of the powder samples do not increase further, even at temperatures above 500 °C, despite the presence of non-oxidized silicon. Due to the much lower thickness of the coatings compared to the radius of the particles, the diffusion path of oxygen is considerably smaller in the coatings, thus a larger mass increase is observed.

After cutting off the upper uncoated area and the lower defect region, samples were subjected to an additional thermal treatment at 1000 °C for 1 h. During this second treatment, the uncoated substrate undergoes an average mass gain more than twice as large as that of the PHPS-coated samples, proving also quantitatively the protective effect of the PHPS-derived SiON coatings, even at high temperatures. It is important to mention that the coatings still contain nitrogen, even after pyrolysis at 1000 °C for 1 h in air, as shown by Günthner et al. [GKDD2009] by means of glow discharge optical emission spectroscopy (GDOES) on coatings with similar thickness. Hence, the mass gain measured after the second pyrolysis is a combination of substrate oxidation, evidenced by the color change and loss of gloss, with further oxidation of the coating.

## 4.2.2   Top-coat

As previously mentioned, different compositions with varied amounts of precursor and of passive and active fillers were investigated. Preliminary tests carried out with precursor amounts varying between 10 and 30 vol% – considering only the volume fractions of precursor and fillers – revealed that a silazane amount below 27 vol% leads to fragile coatings with low adhesion and cohesion. Hence, further tests were performed with silazane amounts of 27 vol%. The amount of each filler was varied as well. Volume ratios of Durazane 1800:ZrSi$_2$ ranging from 1.2:1 to 4:1 with passive filler amount varying from 45 to 80 vol% (related to volume fractions of precursor and fillers) were investigated. It is important to mention that, to obtain low thermal conductivity and high thermal expansion, the amount of passive filler 3YSZ should be as high as possible. Although more than one composition resulted in stable coatings, they possessed different morphologies and critical coating thicknesses. Hence, to select the final composition, these characteristics were also taken into consideration. For the sake of brevity, only two ternary systems are further discussed. For comparison, a system containing only silazane and passive filler and another containing only silazane and active filler were prepared as well. Table 4.2.2 presents the composition and the nomenclature of the four systems discussed herein. The TBC systems were named according to the components present: PP = passive filler + polymer; AP = active filler + polymer; and APP = active filler + passive filler + polymer. As one may notice, the systems TBC-PP, TBC-AP and TBC-APP1 contain 27 vol% of silazane. Thus, by comparing these three systems, the interaction of the two types of filler with the silazane, separately and in combination, can be evaluated. The system TBC-APP2, on the other hand, contains a lower amount of passive filler but the same volume ratio of Durazane 1800:ZrSi$_2$ as in the system TBC-APP1. Thus, a comparison between these two systems offers an insight about the role of the passive filler and of the volume ratio of Durazane 1800:ZrSi$_2$ in the ternary systems.

Table 4.2.2: Composition of the top-coat systems before pyrolysis.

| System | 3YSZ [vol%] | Durazane 1800 [vol%] | ZrSi$_2$ [vol%] | Durazane 1800:ZrSi$_2$ [volume ratio] |
|---|---|---|---|---|
| TBC-PP | 73 | 27 | 0 | - |
| TBC-APP1 | 64 | 27 | 9 | 3:1 |
| TBC-APP2 | 44 | 42 | 14 | 3:1 |
| TBC-AP | 0 | 27 | 73 | 1:2.7 |

*Estimation of composition and CTE after pyrolysis*

Based on XRD, TGA and pycnometry results, the composition of each coating system after pyrolysis was estimated. The volumetric composition of the oxidized ZrSi$_2$

powder – composed of $ZrO_2$, Si, and $SiO_x$ – was determined as explained in section 4.1.2 (see Table 4.1.3). The ceramic yield and the density of the pyrolyzed SiCNO precursor were likewise determined by TGA and helium pycnometry. The volume of 3YSZ is assumed to remain constant during pyrolysis. Using the CTE of each component, the calculated composition and the simplified rule of mixture, the CTE of the coating systems were determined. The obtained values are shown in Table 4.2.3. A CTE of $3 \times 10^{-6}$ $K^{-1}$ was used for the polymer-derived SiCNO fraction [CMRS2010, GSGW2011], $2.6 \times 10^{-6}$ $K^{-1}$ was used for silicon, $0.5 \times 10^{-6}$ $K^{-1}$ for $SiO_x$, and $11.5 \times 10^{-6}$ $K^{-1}$ for 3YSZ/$ZrO_2$.

Table 4.2.3: Estimated compositions and CTEs of the top-coat systems after pyrolysis at 1000 °C for 1 h in air.

| System | 3YSZ/$ZrO_2$ [vol%] | SiCNO [vol%] | Si [vol%] | $SiO_x$ [vol%] | CTE [$\times 10^{-6}$ $K^{-1}$] |
|---|---|---|---|---|---|
| AISI 441 | - | - | - | - | 13.29 |
| TBC-PP | 87.4 | 12.6 | - | - | 10.43 |
| TBC-APP1 | 75.3 | 11.3 | 3.0 | 10.1 | 9.14 |
| TBC-APP2 | 60.7 | 17.8 | 4.8 | 16.3 | 7.74 |
| TBC-AP | 31.2 | 6.5 | 13.8 | 47.2 | 4.44 |

The high amount of 3YSZ in system TBC-PP results in the highest CTE among all systems. The system TBC-APP1 possess a higher CTE compared to TBC-APP2, and the system TBC-AP has the lowest CTE. Hence, from the obtained values, it is expected that the coatings prepared with the formulations TBC-PP and TBC-APP1 show a better thermal compatibility with the substrate than the coatings prepared with the other two systems. However, the stability of the coatings depends also on the strain tolerance and on the behavior of the system and its components during pyrolysis.

*Evaluation of the top-coats and definition of the final composition*

Coatings were deposited by doctor blade technique using an applicator frame with a gap of 120 µm onto steel sheets with dimensions of 90 mm x 90 mm, that were previously coated with a PHPS bond-coat pyrolyzed at 500 °C for 1 h. After top-coat deposition, the samples were cross-linked at 200 °C for 1 h and the coating thickness was measured using the magnetic inductive method. All coatings were about 40 µm thick. After pyrolysis at 1000 °C for 1 h, the samples were inspected visually regarding cracking and spallation.

Fig. 4.2.3a-d present digital images of coatings after pyrolysis. Coatings prepared using the formulation TBC-PP, which contains only the passive filler and the silazane, led to coatings with severe segmentation cracks (across the thickness) and some spallation (Fig. 4.2.3a). The formulation TBC-AP, containing only the active filler and the silazane, resulted in apparently homogeneous, crack-free coatings (Fig. 4.2.3b). Despite the significantly lower CTE of this system in comparison to that of the substrate, a high expansion of the coating was expected during pyrolysis up to 1000 °C due to the high amount of active filler. Curiously, the sample was curved after pyrolysis, as demonstrated in Fig. 4.2.3b. This curvature results from a combination of CTE mismatch, shrinkage of the silazane and expansion of the active filler. The fact that the coating is on the convex face evidences the generation of compressive stresses in the coatings. Typically, adhesion failure – or buckling – of the coating, as observed in EB-PVD TBCs [EvHu1984, VaGJ2000], the formation of cracks and cohesion failure occur when coatings are under stresses. Given the thickness of the substrate and the curvature radius, the fact that no buckling or cohesion failure has occurred is an evidence of strong adhesion and cohesion of the coatings. Further details on this topic will be discussed in section 4.3.2.

Formulation TBC-APP1 yields the best coatings (Fig. 4.2.3c), with no visible cracks and homogeneous appearance. In contrast, coatings prepared with formulation TBC-APP2 (Fig. 4.2.3d) failed completely. Although the same volume ratio of precursor to active filler was used in both compositions, the system TBC-APP2 did not resist the stresses generated during pyrolysis and/or cooling. Thus, the amount of passive filler has proven to be decisive to the stability of the coatings. It influences especially the CTE, which in the case of TBC-APP2 was significantly lower (see Table 4.2.3), increasing the mismatch to the substrate's CTE and, consequently, the stresses during the thermal treatment. Indeed, the coatings remained adhered to the substrate at the borders, where the coatings are thinner, indicating that the critical coating thickness of this system is lower than that of the system TBC-APP1. As observed in the TG analyses, the oxidation of the active filler begins only at ~450 °C. Thus, the shrinkage of the precursor, which constitutes a higher volume fraction in the system TBC-APP2, may have caused excessive shrinkage at lower temperatures. The fact that coatings prepared with the system TBC-AP do not fail but the ones prepared from system TBC-APP2 do is related to the distribution of stresses. In coatings from system TBC-AP, the expansion of the coating occurs homogeneously, due to the presence of only $ZrSi_2$ with the silazane binder phase. In coatings from the system TBC-APP2, the presence of the passive filler causes localized mismatches of expansion within the system, leading to an uneven distribution of stresses, and thus the formation of cracks in the $SiCNO/3YSZ$ matrix. Because the adhesion and cohesion strength of coatings prepared with formulation TBC-APP2 were not sufficiently high, the stresses resulted in coating failure rather than deformation of the substrate. In system TBC-APP1, the lower amount of active filler enables an accommodation of the generated stresses within the $SiCNO/3YSZ$ matrix and the coatings resist the thermal treatment without failing.

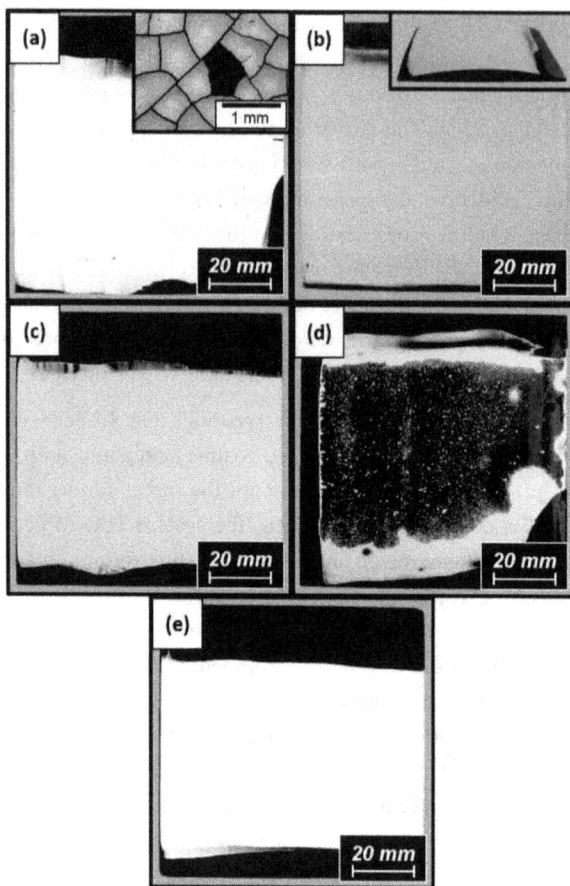

Fig. 4.2.3: Coatings after pyrolysis at 1000 °C in air for 1 h: (a) TBC-PP, (b) TBC-AP, (c) TBC-APP1, (d) TBC-APP2. (e) TBC-APP1 after post-treatment at 1000 °C for 10 h [BaKM2015]. Copyright © 2015 Elsevier Ltd. Adapted with permission.

In order to verify the stability of the coatings prepared with formulation TBC-APP1, a sample was additionally subjected to a thermal treatment at 1000 °C for 10 h. During this post-treatment, the oxidation of $ZrSi_2$ proceeds and reaches its maximum value, according to TGA measurements (see Fig. 4.1.4b). The further oxidation of the active filler is the cause of the color change from a light gray to white, which is characteristic of $ZrO_2$, whereas $ZrSi_2$ and silicon have dark grey color. The progress of the oxidation does not seem to affect the stability of the coatings.

Based on these results, composition TBC-APP1 was selected for further characterizations. The viscosity of the coating suspensions was experimentally adapted to spray coating, in order to enable deposition onto the inside of pipes.

## 4.3 Conversion behavior of the TBC system

Phenomena taking place during pyrolysis are of utmost importance to the stability of the coatings. In order to understand this behavior, mass and dimensional changes were characterized by TGA and dilatometry techniques.

### 4.3.1 Mass change

The conversion behavior of the top-coat system TBC-APP1 was characterized by TGA. For comparison, curves of pure ZrSi₂ and Durazane 1800 with 3 wt% DCP were plotted as well.

The aim of this analysis was to evaluate synergistic effects between the components during thermal treatment. These measurements were performed up to 1000 °C with heating rate of 3 K min⁻¹ in air. In order to eliminate the interference of the dispersant in the analysis, enabling a direct comparison with the curves of individual components, a suspension was prepared following method 2 presented in section 3.2, however excluding the dispersant. The solvent was removed under reduced pressure at room temperature and the obtained powder was then milled with pestle and mortar.

The thermogravimetric behavior of the system TBC-APP1 (Fig. 4.3.1) is consistent with the combination of the curves of 3YSZ, ZrSi₂ and Durazane 1800 with 3 wt% DCP. A negligible mass loss of less than 1% occurs up ~500 °C. Above this temperature, the oxidation of the active filler leads to a mass increase, which at 1000 °C amounts to approx. 4%. The measured mass changes are in very good agreement with calculated values based on composition and mass change of the individual components, indicating that no significant interactions between fillers and precursor occur within the investigated temperature range.

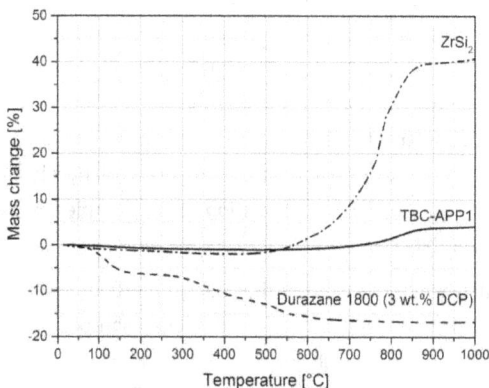

Fig. 4.3.1: *Thermogravimetric behavior of the coating system TBC-APP1 (dried suspension) up to 1000 °C in air with 3 K min⁻¹ in comparison to pure ZrSi₂ and Durazane 1800 with 3 wt% DCP.*

### 4.3.2  Dimensional changes

Dilatometry was applied to measure changes in dimensions of monolithic samples composed of system TBC-APP1 during thermal treatment. Warm-pressed samples prepared from dry coating material were subjected to pyrolysis at 1000 °C for 1 h during dilatometry analyses (Fig. 4.3.2a). During the heat-up phase of the pyrolysis program, the coating material undergoes not only thermal expansion but also dimensional changes caused by precursor shrinkage and active filler expansion. However, the CTE of the system – and thus also the contribution of thermal expansion to dimensional changes during pyrolysis – reduces with the progress of the oxidation of silazane and active filler, due to formation of silicon oxides with low CTE.

A slight expansion takes place in the first 40 min (up to ~235 °C) of the temperature program, which is related to thermal expansion of the components. It is important to mention that the analyzed samples were thermally treated at 160 °C for 2 h during warm pressing. Therefore, the real dimensional changes occurring in low temperature range may differ from those measured during the dilatometry investigations. After this initial expansion, the coating material shrinks in the temperature range of 235-560 °C, due to mass loss and densification of the silazane. Indeed, this shrinkage occurs in two steps, which occur at similar temperatures as the second and third mass loss steps observed in the TGA curve of Durazane 1800 (see Fig. 4.3.1). From 560 °C up to 620 °C, the length of the sample remains virtually constant, evidencing that the shrinkage of the silazane and the expansion of the active filler – which begins at about 450 °C – reach an equilibrium. According to TGA results, the conversion of the active filler amounts only to about 7% up to 620 °C. Above this temperature, an abrupt expansion of the system takes place, which ceases at about 900 °C. TGA investigations have also shown that the oxidation of $ZrSi_2$ takes place in two phases – from 450 to 900 °C and above 900 °C, the first associated to the concurrent oxidation of Zr and Si, and the second related to the oxidation of residual Si. At 900 °C the conversion of the active filler reaches 62%. From 900 to 1000 °C, the length of the sample increases only slightly, which can be attributed to the second phase of the oxidation mechanism of $ZrSi_2$, and to further thermal expansion of the components. However, during the isothermal phase of pyrolysis, no further thermal expansion takes place, whereas the dilatation of the sample can be attributed solely to further expansion of the active filler.

The overall linear expansion of the coating material amounts to 12.3%. Based on composition and TGA results, a completely dense sample of the system TBC-APP1 should undergo a volumetric shrinkage of about 8%, whereas the linear thermal expansion should be lower than 1%, based on the estimated CTE of the coating material. This is attributed to the different temperature ranges where the expansion of the active filler and the hardening of the silazane take place. Durazane 1800 becomes a thermoset polymer at temperatures as low as 130 °C. However, the oxidation of the active filler begins only at ~450 °C. Thus, the expansion of active filler particles cannot be accommodate by the surrounding SiCNO/3YSZ coating matrix by viscous flow. This expansion then results in formation of cracks, leading to a larger expansion of the monolithic samples than expected for the coating material itself.

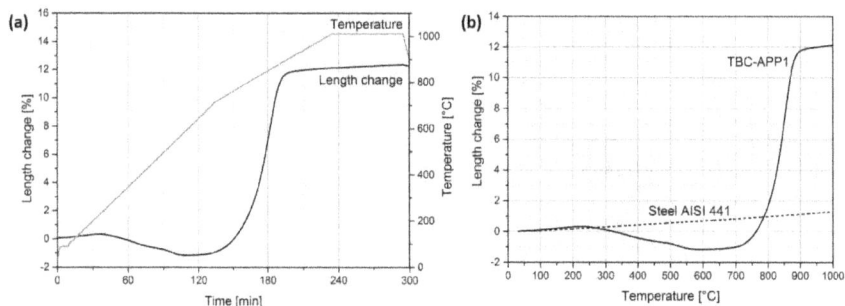

*Fig. 4.3.2: Dilatometry curves in air up to 1000 °C: (a) of a warm-pressed sample prepared from powder of system TBC-APP1 during pyrolysis; and (b) comparison of the expansion of steel 441 and system TBC-APP1.*

Fig. 4.3.2b presents a comparison between the expansions of the steel 441 and the coating material. Up to 250 °C, both substrate and coating material expand equally. Above 250 °C, however, the coating system undergoes shrinkage while the substrate continues to expand. Due to the oxidation of the active filler beginning at 450 °C, the coating material starts to expand and the overall expansions of substrate and coating material become equal at ~785 °C. However, above this temperature the coating material expands further, outmatching the substrate's thermal expansion, resulting in a large expansion mismatch up to 1000 °C. It is important to mention, however, that the relationship between temperature and dimensional changes of warm-pressed samples and coatings may differ due to the different dimensions and microstructures of the samples, which may influence the conversion of the active filler and of the silazane, as well as the accommodation of stresses.

Additionally, warm-pressed powder samples previously pyrolyzed in a conventional furnace were subjected to isothermal prolonged exposure (15 h) at 1000 °C in air during dilatometry (Fig. 4.3.3). These analyses were designed to investigate the effects of further expansion of the active filler after pyrolysis. The expansion at 0 h is the expansion occurring during the heat-up phase of the dilatometry program. Although slightly, the coating material continues to expand during the prolonged isothermal treatment at 1000 °C in air. At constant temperature, no thermal expansion occurs and a further shrinkage of the previously pyrolyzed silazane-derived SiCNO ceramic is unlikely. Thus, the observed dimensional changes result from expansion of the active filler, in agreement with TGA results. The expansion of the system proceeds practically linearly up to approx. 4 h. After that, the expansion rate starts gradually to decrease. After about 10 h, the expansion rate reduces drastically and ceases completely after 12 h, indicating that the maximum conversion of $ZrSi_2$ at this temperature has been reached.

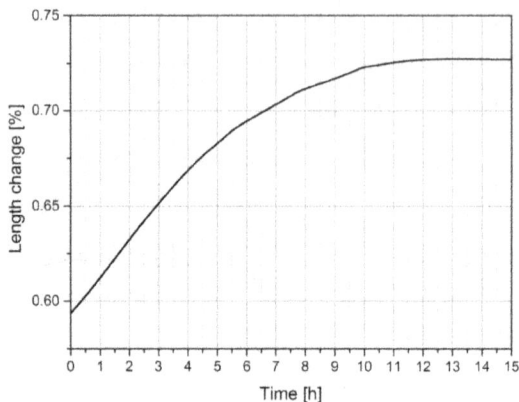

*Fig. 4.3.3: Dilatometry of pre-pyrolyzed, warm-pressed samples prepared from powder of system TBC-APP1 during isothermal treatment at 1000 °C for 15 h in air.*

As observed with system TBC-AP, shrinkage of the silazane, CTE mismatch and expansion of the active filler cause stresses in the systems, which may be relaxed by deformation of the substrate – if adhesion and cohesion of the coating are strong enough – or by cracking and spallation. The coatings prepared with system TBC-APP1 undergo neither spallation nor deformation of the 1.5 mm thick steel sheets after pyrolysis and cooling. This indicates that, despite the high expansion of the coating material revealed by the dilatometry investigations, stresses arising during pyrolysis and cooling are not sufficient to overcome adhesion and cohesion forces nor the stiffness of the substrate. However, bending of the substrate is observed after pyrolysis of coatings on steel sheets with thickness of 0.20 mm, with the coating on the convex face, similarly to the coatings prepared with formulation TBC-AP. This proves that also coatings of system TBC-APP1 are under compressive stresses after pyrolysis and cooling. While the thermal expansion is a reversible phenomenon, neither the expansion of the active filler nor the shrinkage of the precursor are reversed upon cooling after pyrolysis. Due to active filler expansion and the higher CTE of the substrate, the presence of residual compressive stresses in the coatings was expected after cooling. Since both adhesion and cohesion of the coating are strong, stresses in the system are relieved by bending the substrate.

Although the presence of compressive stresses after cooling is evident, the evolution of the stresses during pyrolysis is still undetermined. There are three different scenarios, which may describe this evolution (Fig. 4.3.4):

- Scenario 1: the effects of substrate's CTE and of precursor shrinkage outmatch the effects of active filler expansion and coating thermal expansion. As a result, the substrate expands more than the coating, curving the coated sheet with the coating on the concave face. During cooling, the substrate contracts more than the coating due to the CTE mismatch, reversing the curvature, whereas the coating now lies on the convex side.

- Scenario 2: the mismatch between dimensional changes of coating and substrate are not sufficient to overcome the stiffness of the substrate, whereas the coated sheet remains flat until the cooling phase begins. During the cooling phase, similarly to the scenario 1, the CTE mismatch causes a higher contraction of the substrate compared to the coating, leading to the bending of the sheet, with the coating on the convex side.

- Scenario 3: the expansion of the active filler and the thermal expansion of the coating material cause and overall expansion of the coating, which outmatches the thermal expansion and the stiffness of the substrate, resulting in the bending of the coated sheet, with the coating on the convex side, even at high temperatures. The curvature radius then decreases further during cooling, due to the CTE mismatch.

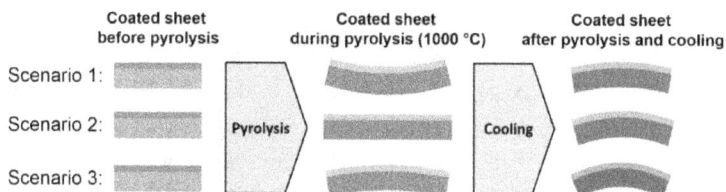

Fig. 4.3.4: Scenarios of the curvature evolution during pyrolysis of the PDC-based TBC.

A thermo-optical analysis of coated samples enables the evaluation of the curvature evolution. Measurements were carried out during the pyrolysis program and during a prolonged isothermal treatment at 1000 °C. As already mentioned, coatings had to be pre-pyrolyzed at 700 °C, to protect the equipment from gaseous products of pyrolysis reactions. Fig. 4.3.5 displays the curvature evolution and digital photographs taken during the pyrolysis program (up to 700 °C with heating rate of 5 K min⁻¹ followed by a second heating phase up to 1000 °C with 3 K min⁻¹, and an isothermal phase of 1 h). After pyrolysis, the furnace was shut down and the sample was cooled naturally inside the furnace. The capital letters at the lower left corner of each photograph in Fig. 4.3.5

correspond to the points marked on the graph and identify the moment, at which each photograph was taken.

Due to the pre-pyrolysis at 700 °C, an almost negligible curvature, with the coating on the convex side of the metal sheet, is observed prior to the thermo-optical dilatometry investigations (A in Fig. 4.3.5). This suggests that a slight expansion mismatch between coating and substrate occurs already during the pre-treatment up to 700 °C in air. During the cooling phase after the pre-treatment, the thermal expansion of the substrate is reversed, resulting in the observed curvature. However, this curvature decreases as the temperature rises during the thermo-optical dilatometry measurements, due to the higher CTE of the substrate compared to that of the coating, whereas at about 700 °C the sheet is flat. This indicates that the coating expansion, which occurred during the pre-treatment, was not sufficient to overcome the thermal expansion of the substrate. Between 700 and 820 °C, the curvature increases again, but starts to decrease as the temperature continues to rise, whereby the sheet becomes flat at 1000 °C in the end of the heat-up phase of the pyrolysis program (B in Fig. 4.3.5).

Based on results of conventional dilatometry investigations, which revealed a much larger expansion of the coating material than of the substrate up to 1000 °C (see Fig. 4.3.2b), a curvature with the coating on the convex side was expected. As observed by conventional dilatometry of monolithic samples, up to 550 °C the coating material shrinks while the substrate expands. In order to cause an overall expansion of the coating, the expansion of the active filler must firstly outmatch the shrinkage of the precursor. A higher expansion of the coating material compared to the substrate is only observed above ~800 °C. In coated sheets, the initial shrinkage of the coating material is constrained by the substrate, generating tensile stresses within the coatings, which may result in segmentation cracks, as is usually observed after sintering of ceramic films [BoRa1985, BoJa1993]. Thus, the initial expansion of the active filler occurring above 450 °C is accommodated within the coating, reducing the tensile stresses without causing any expansion of the coating. With further filler expansion, the stresses eventually become compressive. Due to the much shorter diffusion paths, higher oxidation rates are expected in coatings compared to monolithic samples. Thus, the expansion of the active filler occurs faster within coatings. Indeed, at 700 °C the compressive stresses become high enough to overcome the thermal expansion and the stiffness of the substrate, causing a curvature increase between 700 and 820 °C. However, due to a further increase of substrate's CTE (see Fig. 4.1.3) and a reduction of the oxidation rate of $ZrSi_2$ (see Fig. 4.1.4a), above 820 °C the substrate starts to expand more than the coating again, and the metal sheet becomes flat at 1000 °C.

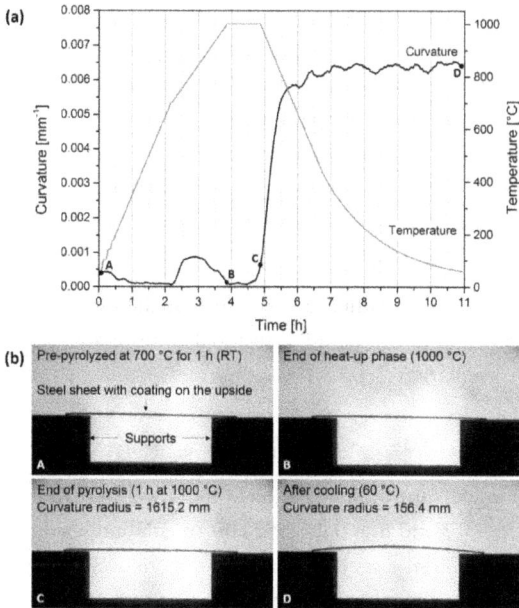

Fig. 4.3.5: Thermo-optical dilatometry investigations during pyrolysis at 1000 °C in air for 1 h of a coating deposited by spraying with formulation TBC-APP1 onto a 0.20 mm thick steel 441 sheet: (a) curvature evolution; (b) digital photographs and curvature radii (A = after pre-pyrolysis at 700 °C for 1 h; B = end of heat-up phase; C = end of isothermal phase; D = end of cooling phase).

Once the metal sheet reaches the final pyrolysis temperature of 1000 °C, the thermal expansion of the substrate ceases. Interestingly, despite the continuation of the oxidation reactions of the active filler (see Fig. 4.1.4b), the metal sheet remains flat up to 40 min at 1000 °C. Only during the last 20 min of the isothermal phase, the curvature increases (C in Fig. 4.3.5). Although until this point the curvature radius is still very large (1615.2 mm), the bending of the substrate with the coating on the convex face could be observed even before cooling, evidencing the occurrence of scenario 3. This indicates that the coating expansion is larger than the thermal expansion of the substrate. Thus, the PDC-based TBC is under compressive stresses, even at 1000 °C.

During the cooling phase after pyrolysis, the thermal expansion is reversed. Since the CTE of the coating is lower than that of the substrate and the filler expansion is irreversible, the substrate retracts more than the coating. Thus, the compressive stresses in the coatings increase as the temperature decreases. During the cooling phase, the curvature increases more abruptly between 1000 and ~700 °C, and more slowly afterwards. Below 400 °C, the curvature remains constant. This is attributed to the CTE evolution of the uncoated steel (see Fig. 4.1.3), which changes more abruptly at the high temperature range (above 700 °C). As the linear CTE of the steel diminishes, also the mismatch with the CTE of the coating decreases and thus, a less significant curvature change is observed. Also the expansion of the active filler proceeds during the cooling phase, as long as sufficient temperature for the oxidation reactions is provided. Moreover, the Young's modulus of the substrate increases as the temperature decreases. With the higher Young's modulus increases the resistance of the substrate to deformation. At the end of the cooling phase of pyrolysis (D in Fig. 4.3.5), the radius of curvature amounts to 156.4 mm.

Using the measured radius of curvature at the end of the cooling phase after pyrolysis, it is possible to estimate the residual compressive stress within the coating using the Stoney's equation (Eq. 3.5.11).

$$\sigma_{St} = \frac{E_s d_s^2}{6R_\kappa(1 - v_s)d_c}$$                  (Eq. 3.5.11)

Input parameters for the calculation of the stress were $E_s$ = 220 GPa, $d_s$ = 0.2 mm, $R_\kappa$ = 156.4 mm, $v_s$ = 0.3 and $d_c$ = 0.03 mm. This calculation results in a residual stress of approx. 0.45 GPa within the coating at room temperature, which is significantly lower than typical stresses of about ~1 GPa induced by the growth of the TGO in conventional TBCs [SSBP2000].

In order to determine the maximum compressive stress arising within the PDC-based TBCs, a post-treatment was carried out after pyrolysis (up to 1000 °C with a heating rate of 5 K min⁻¹, followed by an isothermal phase of 10 h, and free cooling). The curvature evolution during this post-treatment was characterized by thermo-optical dilatometry as well (Fig. 4.3.6).

During the heat-up phase of the post-treatment, the metal expands more than the coating due to the CTE mismatch, resulting in an increase of the curvature radius from 156.4 mm before heating to 218.9 mm at the end of the heat-up phase (1000 °C), i.e. the sheet is less curved (E in Fig. 4.3.6). This evidences a relaxation of the system. At this point, the overall stress amounts to only 0.15 GPa according to Stoney's equation (with $E_s$ = 100 GPa, $d_s$ = 0.2 mm, $R_\kappa$ = 218.9 mm, $v_s$ = 0.3 and $d_c$ = 0.03 mm).

However, as expected based on the TGA and dilatometry results, the post-treatment revealed that the expansion of the coating continues after pyrolysis due to further oxidation of the active filler at high temperatures. Coating expansion takes place at the largest rate up to ~4 h at 1000 °C. The curvature radius decreases from 218.9 mm at the end of the heat-up phase, to only 48.8 mm after 4 h at 1000 °C (G in Fig. 4.3.6). From there on, the rate decreases but the expansion of the coating proceeds, with consequent further decrease of the curvature radius. The curvature stabilizes after 9-10 h at 1000 °C, indicating that oxidation reactions of the active filler have stopped. After 10 h at 1000 °C, the radius of curvature of the sheet amounts to 30.2 mm (J in Fig. 4.3.6). This curvature corresponds to a compressive stress of 1.05 GPa, according to Stoney's equation (with $E_s$ = 100 GPa, $d_s$ = 0.2 mm, $R_\kappa$ = 30.2 mm, $v_s$ = 0.3 and $d_c$ = 0.03 mm), which is similar to the typical stresses induced by the growth of the TGO in conventional TBCs [SSBP2000].

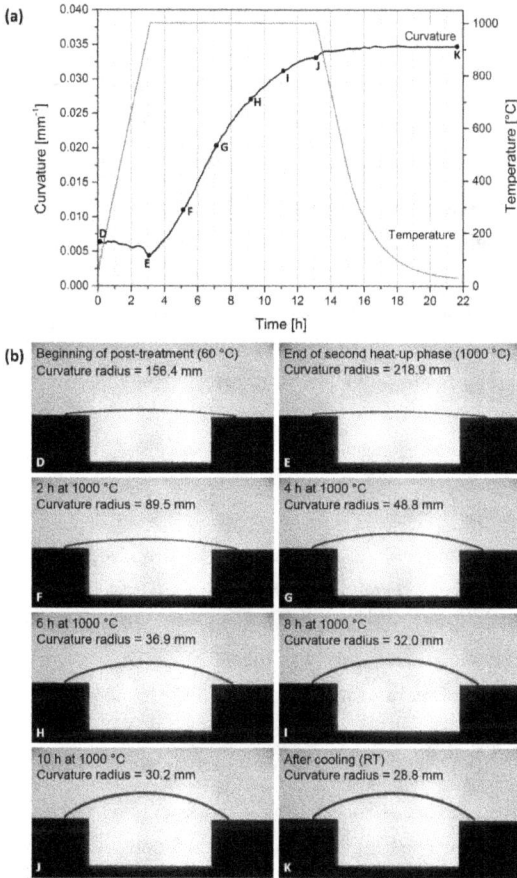

Fig. 4.3.6: *Thermo-optical dilatometry investigations during post-treatment at 1000 °C in air for up to 10 h of a pyrolyzed coating deposited by spraying with formulation TBC-APP1 onto a 0.20 mm thick steel 441 sheet: (a) curvature evolution; (b) digital images and curvature radii (D = before post-treatment; E = end of heat-up phase; F-I = during the isothermal phase; J = end of isothermal phase; K = end of cooling phase).*

During cooling, the curvature increases more significantly at 1000-900 °C. Below ~300 °C the curvature does not change anymore. After cooling down to room temperature, the radius of curvature is 28.8 mm (K in Fig. 4.3.6). However, the increase of the curvature during the cooling phase after the post-treatment is significantly lower than that observed during the cooling phase after pyrolysis. This might be related to the fact that no further expansion of the active filler takes place during the cooling phase after pyrolysis.

Since the expansion of the active filler is already complete, the maximum value of compressive stress is reached when the system is cooled down to room temperature, resulting in an overall stress of 2.43 GPa, as calculated using Stoney's equation (with $E_s$ = 220 GPa, $d_s$ = 0.2 mm, $R_\kappa$ = 28.8 mm, $v_s$ = 0.3 and $d_c$ = 0.03 mm). This value is still significantly inferior compared to the overall residual stresses of up to ~7 GPa, which may occur in conventional TBCs [LiCl1996], caused mostly by CTE mismatch.

The permanent state of compression of the PDC-based TBCs prevents the formation of segmentation cracks, which could increase the thermal conductivity of the system. Moreover, creep of the substrate is likely to occur upon long-term exposure of the metal to high temperatures. While tensile stresses would arise in stress-free coatings, which could result in coating failure, the presence of compressive stresses may increase the resistance of the system to the effects of creep.

## 4.4    Microstructure

### 4.4.1    Morphology of the cross-section

Coatings with different thicknesses were prepared by varying the deposition parameters in order to determine the maximum coating thickness of the system TBC-APP1. The cross-section of the intact coatings with the largest thickness according to magnetic induction measurements was then further analyzed by SEM (Fig. 4.4.1). The images evidence the preparation of coatings with thickness of about 50 µm. The PHPS-derived bond-coat is about 1 µm thick, as expected based on the data published by Günthner et al. [GKDD2009], but varies due to the roughness of the substrate.

The microstructure of the coating's cross-section is characterized by extensive cracking. However, the cracking pattern contains mostly horizontal and diagonal cracks – as shown in red and blue, respectively, in Fig. 4.4.1b – with some small vertical cracks, which do not coalesce to form segmentation cracks. As discussed in section 2.3.2, the obtained cracking pattern is the ideal microstructure to obtain low thermal conductivity [PaGJ2002].

The absence of segmentation cracks is attributed to the compressive stresses arising within the coatings, as discussed in the previous section. Furthermore, also the presence of diagonal cracks results from compressive stresses. According to Brace and Bombolakis [BrBo1963], the most intensely stressed cracks, thus the ones which will most likely propagate upon the application of compressive forces, are those inclined 30° to the direction of the force. As shown in Fig. 4.4.1b, this is virtually the same inclination of most of the diagonal cracks marked in blue. In contrast, according to the same authors, upon a tension applied uniaxially, the most stressed cracks are those perpendicular to the tensile force. Thus, the presence of such cracks, marked in red in Fig. 4.4.1b, suggest the occurrence also of tensile forces normal to the coating surface during thermal treatment. Such tensile forces may arise due to the expansion of the coating, which is unconstrained only in this direction, resulting in an increase of coating thickness.

Fig. 4.4.1c presents a micrograph of the filler particles within the coatings. As expected based on the particles size distribution, the active filler particles ($D_{90}$ = 3.0 μm) are easily identified among the submicron 3YSZ particles ($D_{90}$ = 0.5 μm) in high-magnification images. Moreover, the heterogeneous morphology of the active filler particles after pyrolysis, caused by the formation of two different oxides – oxides of zirconium and silicon – and free silicon becomes evident. In this image, the presence of some residual porosity in the interparticle spaces is observed. As intended, the silazane acts only as binder for the filler particles, whereas the formation of a SiCNO matrix is not observed.

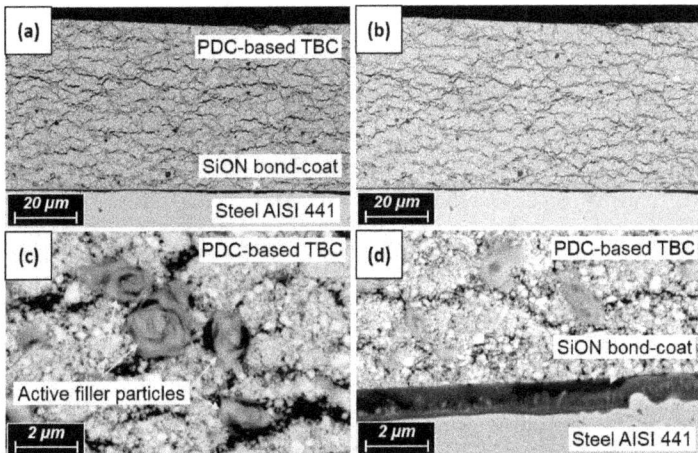

*Fig. 4.4.1: SEM micrographs of the cross-section of a coating deposited onto steel 441 by spraying with formulation TBC-APP1 followed by pyrolysis at 1000 °C for 1 h in air: (a) overview; (b) highlight of the cracking pattern: red for horizontal and blue for diagonal cracks; (c) high-magnification image of the coating; (d) high-magnification image of the bond-coat region.*

Another interesting feature of the cross-section microstructure refers to the interaction of the substrate with the PHPS-derived bond-coat. Fig. 4.4.1d shows a high-magnification SEM micrograph of the bond-coat region. There seems to occur some diffusion of elements from the stainless steel into the SiON bond-coat, forming protrusions within the bond-coat, which grow toward the top-coat. Similar findings were reported by Schütz et al. [SGMG2012], also for the case of silazane-derived coatings on steel substrate.

The formation of such protrusions has been described by Issartel and colleagues [IMCP2012] upon oxidation of uncoated ferritic stainless steel 441 at 1060 °C. They proposed an oxidation mechanism composed of four phases. In the first phase, the metal/oxide interface becomes undulated due to an inhomogeneous growth of the oxide scale. This induces an increase of the concentration of oxygen in the troughs of the undulations. In the second phase, observed already during the first seconds of oxidation of uncoated sheets at 1060 °C, the higher oxygen content in the troughs induces nucleation of silicon oxides and causes a faster advance of the oxidation front toward the interior of the metal in these regions – increasing the amplitude of the undulations, forming protrusions. These protrusions are composed mostly of Fe, Cr, and oxides of Si and Nb. In the third phase, observed after about 15 min at 1060 °C, the silicon oxide layer grows, consuming the bridges of the protrusions and creating metallic inclusions within the Cr, Mn, Fe oxide scale. In the fourth phase, after 30 min of exposure, the inclusions oxidize and become part of the oxide scale, whereas the silica film remains at the interface.

This thin silica layer is held responsible for the adhesion failure of the oxide scale, due to its low CTE [JaCS2010, GrFS2014]. However, the formation of a silicon oxide layer is not always observed, which might be a consequence of the capture of silicon by the $Fe_2Nb$ Laves phase formed in the steel [JaCS2010, GrFS2014]. Temperature, time and niobium content are factors influencing the formation of the silicon oxide layer.

In the case of PHPS-coated samples, however, the SiON layer exists prior to the formation of the oxide scale and protrusions seem to form by diffusion of metallic elements through the bond-coat. EDS element mappings (Fig. 4.4.2) provide evidence of the diffusion of Cr and Mn through the bond-coat, whereas Ti initially segregates near the interface with the bond-coat. No diffusion of iron toward the coating occurs during pyrolysis. Bond-coats of conventional TBC systems are designed to slowly oxidize and generate a TGO of $Al_2O_3$, capable of blocking further penetration of oxygen toward the substrate. In contrast, the PHPS bond-coat oxidizes quickly during the first pyrolysis and acts like an oxygen barrier from there on, blocking the diffusion of oxygen toward the substrate, similarly to a TGO. However, the EDS mappings revealed that, although the PHPS-derived bond-coat is able to avoid the formation of an oxide scale directly onto the substrate, by preventing the inward diffusion of oxygen, it is not able to avoid the outward diffusion of metallic elements. The inhomogeneous diffusion of these elements, attributed to the somewhat rough surface of the metal, may be the cause of the formation of the protrusions and inclusions.

Fig. 4.4.2: SEM micrographs with EDS element mappings of the bond-coat region of a coating deposited onto steel 441 by spraying with formulation TBC-APP1 followed by pyrolysis at 1000 °C in air for 1 h.

EDS was also used to investigate the distribution of filler particles within the top-coat (Fig. 4.4.3). As expected, zirconium is distributed homogeneously, which is a consequence of the presence of Zr in both fillers and the large amount of passive filler. Silicon, on the other hand, is not homogeneously distributed within the coating. Although this element is present across all the coating, due to the polymer-derived SiCNO binder phase, it is more concentrated in some spots. Some of these spots match with dark gray spots in the SEM images, which appear to be pores filled with the silazane. The other silicon agglomerates correspond to the active filler particles, which are homogeneously distributed as well.

Fig. 4.4.3: SEM micrographs with EDS element mappings of the cross-section of a coating deposited onto PHPS-coated steel 441 by spraying with formulation TBC-APP1 followed by pyrolysis at 1000 °C in air for 1 h.

The identification of the active filler particles enables a correlation of the active filler expansion with the cracking pattern. In Fig. 4.4.4, the position of silicon spots on the EDS mapping were marked with green dots on the SEM micrograph to facilitate visualization. As it can be observed, the cracking pattern is related to the active filler particles, with several cracks starting from the particles toward the SiCNO/3YSZ matrix, as shown in the zoomed images in Fig. 4.4.4. This is explained by the fact that the active filler particles continue to expand even after hardening of the precursor and consolidation of the SiCNO/3YSZ matrix. This is in good agreement with results of the dilatometry investigations. The expansion of the particles generates microstresses in the SiCNO/3YSZ matrix, which are compressive in the radial coordinate and tensile in the polar and azimuthal coordinates of the spherical system. The intensity of these stresses decrease at a rate of $r^{-3}$, where $r$ is the distance to the center of the particle, if the particles are somewhat isolated within the matrix [Todd2006]. This combination of stresses cause the formation of cracks along the radial coordinate, as schematically shown in Fig. 4.4.5. These cracks then propagate within the SiCNO/3YSZ matrix and coalesce with other cracks in the vicinity. Thus, both diagonal and horizontal cracks may be explained by the expansion of the active filler. In the direction parallel to the substrate, the expansion of the active filler generates compressive stresses due to the constrained geometry, analogously to the constrained sintering phenomenon. In the direction perpendicular to the substrate, on the other hand, the expansion of the active filler particles is constrained only by the surrounding matrix, leading to an expansion of the coating thickness, thus generating overall tensile stresses, which induce the formation of horizontal cracks. The fact that the vertical cracks do not coalesce and form segmentation cracks is a consequence of compressive stresses parallel to the substrate, as discussed previously.

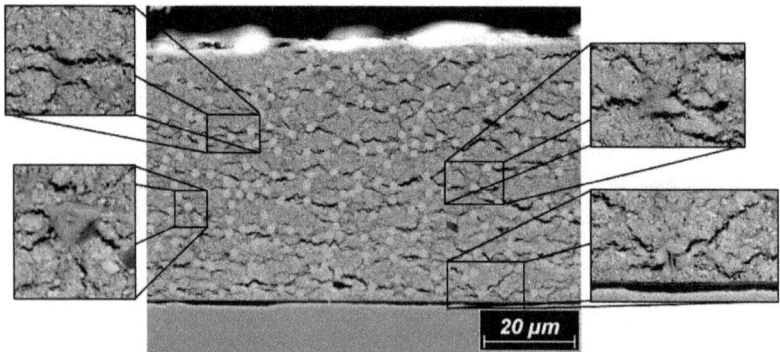

*Fig. 4.4.4: SEM cross-section micrograph of a coating deposited onto steel 441 by spraying with formulation TBC-APP1 followed by pyrolysis at 1000 °C for 1 h in air: correlation between active filler particles and cracking pattern.*

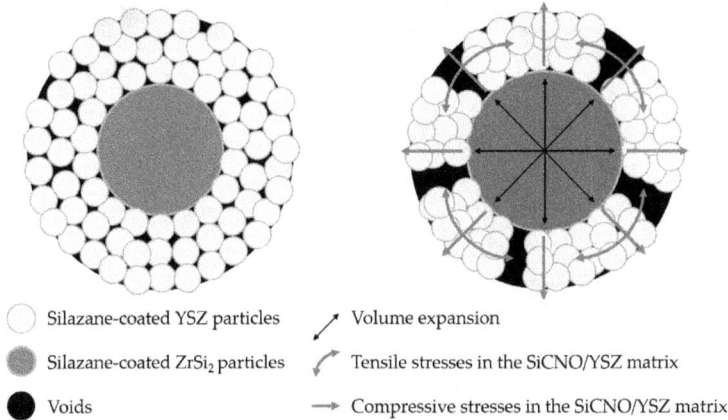

| ○ Silazane-coated YSZ particles | ↗ Volume expansion |
| ● Silazane-coated ZrSi$_2$ particles | ↗ Tensile stresses in the SiCNO/YSZ matrix |
| ● Voids | → Compressive stresses in the SiCNO/YSZ matrix |

*Fig. 4.4.5: Schematic representation of the effect of active filler expansion in the SiCNO/3YSZ coating matrix.*

Coating porosity was evaluated by analysis of binary SEM micrographs using the area fraction measurement tool of the software ImageJ [ScRE2012]. After analysis of 20 images, the obtained porosity values were averaged. Fig. 4.4.6 shows an example of such binary image used for porosity quantification. The image analyses resulted in an average porosity value of 7.9%, with a standard deviation of 1.3%. The presence of particles with different size distributions enables a good packing of the fillers and the liquid state of the preceramic polymer enables the filling of voids in the interparticle spaces. However, the low volume fraction of silazane and the mass loss due to evaporation of solvent, dispersant burn-out and ceramization of the precursor may result in some residual porosity. However, the observed porosity is constituted mainly by cracks and only a reduced amount of residual porosity is present in the interparticle spaces, as observed previously (see Fig. 4.4.1c). A probable cause for this is the expansion of the active filler, as schematically shown in Fig. 4.4.5. This expansion leads simultaneously to formation of cracks and to a compaction of the SiCNO/3YSZ matrix around the active filler particles, closing at least some of the residual porosity. Moreover, compressive stresses in the coatings during cooling may reduce the transient porosity in comparison to the high temperature state.

Porosity plays an ambiguous role in TBC systems. On one hand, a high (closed) porosity reduces the thermal conductivity of the coating, as air has an almost negligible thermal conductivity. Up to a certain amount, the presence of porosity is also beneficial to the mechanical stability of the system, as it increases the strain compliance by reducing the overall Young's modulus of the coatings. On the other hand, an excessive porosity reduces the cohesion strength, whereby coatings may become fragile.

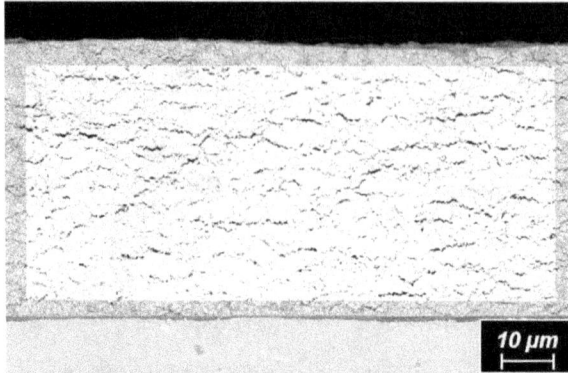

Fig. 4.4.6: Example of binary SEM cross-section micrograph for quantification of porosity. Coating deposited onto steel 441 by spraying with the formulation TBC-APP1 followed by pyrolysis at 1000 °C for 1 h in air.

### 4.4.2 Morphology of the surface

The surface morphology of the top-coat was investigated by SEM (Fig. 4.4.7). The surface is characterized by small cracks originating from active filler particles located near the surface. Due to their proximity to the surface, the expansion of these $ZrSi_2$ particles cause an elevation of the coating directly above, as shown in the zoomed detail of Fig. 4.4.7.

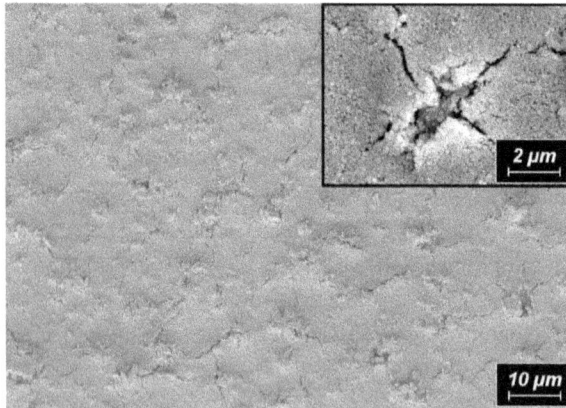

Fig. 4.4.7: SEM micrograph of the surface of a coating deposited onto steel 441 by spraying with formulation TBC-APP1 followed by pyrolysis at 1000 °C for 1 h in air.

### 4.5    Coating adhesion

Adhesion of the coatings was investigated initially by cross-cut tape test (DIN EN ISO 2409 [DIN2013]). A visual inspection using a loupe revealed the occurrence of some spallation. The coatings were classified as class Gt-1 in a scale from Gt-0 (best) to Gt-5 (worst). Class Gt-1 means that up to 5% of the tested coating area was removed by the tape. However, the most interesting information obtained from this test is the failure mechanism. Hence, further analysis was performed by optical microscopy. Fig. 4.5.1a,b present two examples of images obtained using the dark field mode of the optical microscope. In these images, it becomes clear that the predominant failure mechanism is cohesion failure, in which the system fails within the coating, thus not exposing the substrate. Some local adhesion failure has also occurred, as shown in 4.5.1a, but the contribution of this failure mechanism is much less significant. These localized adhesive failures may be related to failures in the bond-coat, as discussed in section 4.2.1. Fig. 4.5.1c,d show optical microscopy images obtained using the differential interference contrast mode. These images enable the quantification of the undamaged area using the software ImageJ. These analyses resulted in an average of 49% of undamaged area, i.e. around 51% of the coated area is affected by cohesion failure. This type of failure mechanism was expected based on the microstructural features of the top-coats, discussed in section 4.2.2.

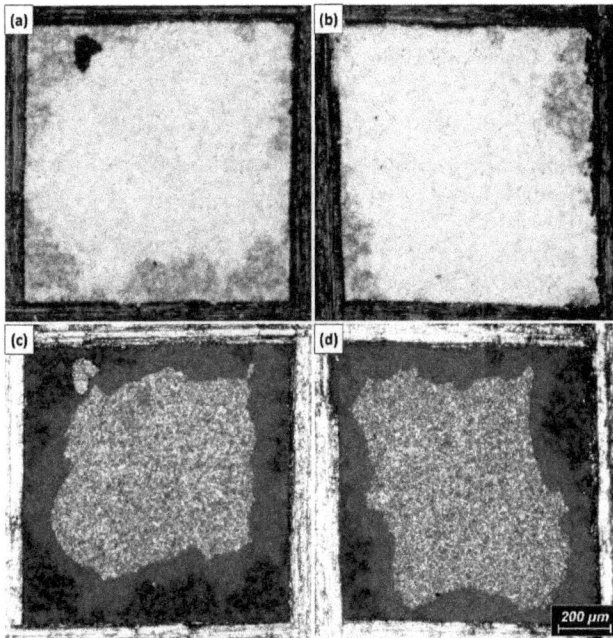

Fig. 4.5.1: Optical microscopy images of a coating deposited onto steel 441 by spraying with formulation TBC-APP1 followed by pyrolysis at 1000 °C for 1 h in air after cross-cut tape test: (a,b) dark field mode, (c,d) differential interference contrast mode.

In order to quantify the adhesion of the coating system, pull-off adhesion tests (ASTM D4541 [ASTM2009]) were performed. The measurements resulted in an average pull-off adhesion of 20.9 MPa (n = 4, $\sigma_{SD}$ = 2.0 MPa). The adhesion of conventional TBCs is generally characterized by tensile tests as well, whereas the most common method is the tensile adhesion test ASTM C633 [ASTM2013]. This method is, however, limited to coatings with thickness of 300 µm or more. Nevertheless, the tensile adhesion of conventional TBCs is usually in the range of 10 to 50 MPa [EBJÖ2011]. Thus, the PDC-based TBCs have comparable adhesion with conventional TBC systems.

The strong adhesion of the coatings arises from the chemical bonding of the silazane with substrate and fillers. According to Amouzou and colleagues [AFMS2014], who investigated the formation of chemical bonds between silazanes and stainless steel substrates by X-ray photoelectron spectroscopy (XPS) and infra-red reflection absorption spectroscopy (IRRAS), N-H bonds in silazanes react with -OH groups at the surface of the substrate and filler particles to form covalent bonds with elimination of ammonia. In addition, due to the presence of humidity adsorbed at the surface of the substrate or from the environment, reactive groups, such as Si-H, can be hydrolyzed, forming silanols (Si-OH), which also interact and condense with –OH groups at the surface of substrates and fillers, forming oxygen bridges and eliminating water in the process [vZPJ2013, PPFG2015]. Compared to other mechanisms, like mechanical interlocking, electrostatic and acid-base interactions, covalent bonding is considered to be the strongest adhesion mechanism [Kinl1987].

To evaluate failure mechanisms, macro digital images of the coating and dolly surfaces were evaluated (Fig. 4.5.2). Failure has not occurred homogeneously throughout the tested area. Damage was concentrated mostly at the center of the tested area. At the outer region, failure seems to have occurred within the resin layer or at the interface with the dolly, as confirmed by the exposed aluminum at the surface of the dolly. Therefore, it is possible to affirm that the measured adhesion is most likely underestimated, since part of the adhesion failure is not related to the coatings, but rather to the resin layer.

Within the damaged area at the coating's surface there are two distinct regions. One characterized by dark brown coloration and the other is gray. The gray region at the coating surface corresponds to a region with dark brown coloration in the analogous position at the dolly's surface. Based on the difference of coloration of uncoated substrates before and after the thermal treatment at 500 °C for 1 h and additionally at 1000 °C for 1 h (see Fig. 4.2.2), one can conclude that the oxide scale was removed from the substrate in the regions with gray coloration, exposing non-oxidized substrate. To better understand how the coatings failed, EDS mappings were performed in part of the tested area (Fig. 4.5.3).

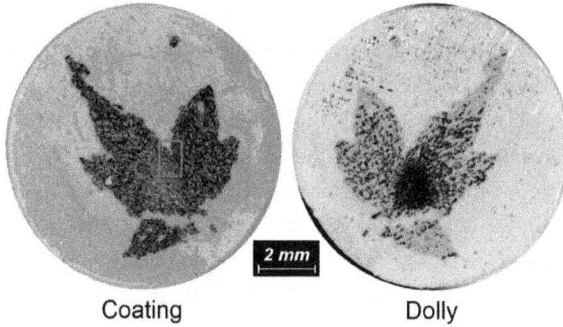

Coating                    Dolly

Fig. 4.5.2: Macro digital images of coating and dolly surfaces after pull-off test (Adhesion of 19.6 MPa). Coating deposited onto PHPS-coated steel 441 by spraying with formulation TBC-APP1 followed by pyrolysis at 1000 °C for 1 h in air. The red rectangle marks the region investigated by EDS (Fig. 4.5.3).

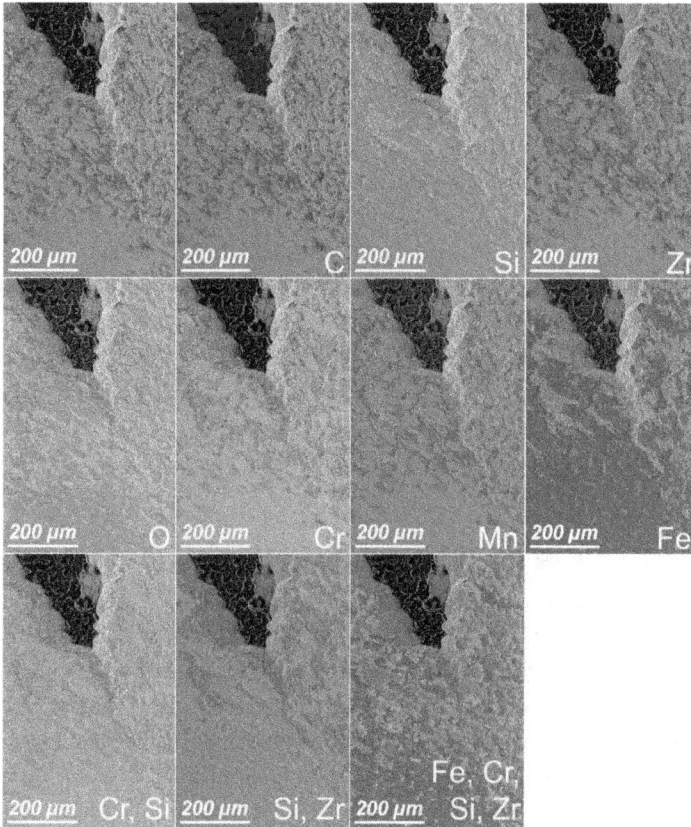

Fig. 4.5.3: EDS mappings on the surface of a coating deposited by spraying with formulation TBC-APP1 onto steel 441 followed by pyrolysis at 1000 °C for 1 h in air after pull-off test.

The occurrence of failure within the resin or at the interface dolly/resin was confirmed by carbon mapping. Mapping of oxygen confirmed the removal of the oxide scale in a considerable fraction of the mapped area. The regions where the color is more intense in the silicon map identify the PHPS-derived SiON layer. Interestingly, the mapping of zirconium revealed that the occurrence of cohesion failure within the top-coat is not the dominant failure mechanism, in contrast to the results of the cross-cut tape tests. This may be attributed to the different stresses applied to the coatings. In the pull-off test, the stresses are tensile and perpendicular to the coating surface, which induces horizontal crack propagation. Despite the crack pattern of the top-coats, little cohesion failure has occurred, attributed to the high cohesion within the coatings. The green regions on the overlapped map of silicon and zirconium identify the areas where the SiON layer became exposed, indicating that either cohesion failure has occurred within the SiON layer or adhesion failure occurred at the interface bond-coat/top-coat. This may be related to the diffusion of the metallic elements through the PHPS-derived layer, reducing its stability. By comparing the maps of chromium and silicon, the presence of the former within the SiON layer becomes evident, as it was observed in the cross-section SEM-EDS analyses discussed in section 4.4.1. The overlapped maps of silicon, zirconium, chromium and iron leads to the conclusion that the failure mechanism of the PDC-based TBC system is complex and involves adhesion and cohesion failure mechanisms in all layers and interfaces, since bare substrate, oxide scale, bond-coat and top-coat are visible after the pull-off tests. However, it seems that the adhesion of the PHPS-derived layer to the substrate is the limiting factor, which is related to the non-ideal surface of the substrate and to the diffusion of metallic elements across the bond-coat. Thus, the improvement of the quality and of the properties of the bond-coat seems to be crucial to enhance the tensile adhesion of the whole coating system.

## 4.6 Thermal properties

### 4.6.1 Coefficient of thermal expansion

As previously discussed, during the first hours of thermal treatment, the composition of the coating system changes, which causes also dimensional changes. Hence, the true linear CTE of the coating can only be obtained after completion of the reactions within the coating material. Dilatometry measurements were carried out on pyrolyzed monolithic samples after a prolonged oxidation treatment of 15 h at 1000 °C in air. After this treatment, the active filler is fully oxidized and the particles do not expand further. Thus, any length change occurring at this point results from thermal expansion/retraction. Measurements in the range of room temperature up to 1000 °C in air, with 10 K min⁻¹ were carried out (Fig. 4.6.1). For comparison, the expansion of the steel and of the same sample directly after pyrolysis are also presented.

The most important information obtained with these analyses is the CTE mismatch between coating material and substrate. Fig. 4.6.2 presents the evolution of the linear CTE of the steel and of the coating system before and after the 15 h long treatment at 1000 °C, as well as the calculated CTE mismatches. While the CTE of the steel increases with increasing temperature, the CTE of the coating material is virtually temperature-independent. As a result, the CTE mismatch between coating and substrate – and consequently also the thermal stresses – increases with increasing temperature. This explains why the curvature of the thin coated metal sheets increases more significantly in the high temperature range during the cooling phase, as discussed in section 4.3.2.

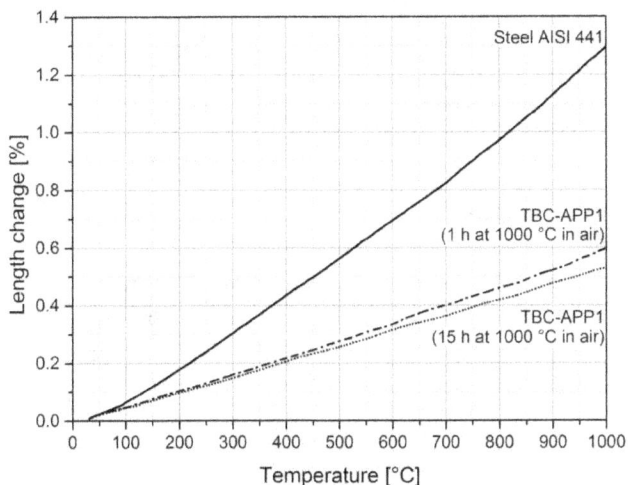

Fig. 4.6.1: Dilatometry curves of warm-pressed samples prepared from powder of formulation TBC-APP1 after thermal treatment at 1000 °C for 1 and 15 h in air in comparison to the steel 441. Measurement in the range of 30 to 1000 °C in air, with 10 K min⁻¹.

Another interesting feature is the decrease of the CTE of the coating material after 15 h of oxidation at 1000 °C. This is attributed to further oxidation of the active filler forming additional $SiO_2$, which reduces the overall CTE of the system. Moreover, before this treatment, the expansion of the material results from the combination of thermal expansion and active filler expansion, whereas after the complete oxidation of the system during the post-treatment, only thermal expansion occurs. However, the fact that the values differ even in the low temperature range, where filler oxidation does not take place, indicates that the decrease of the CTE is mainly caused by changes in composition.

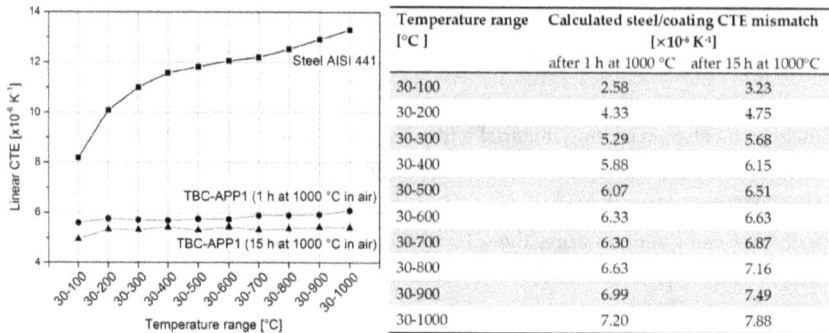

| Temperature range [°C] | Calculated steel/coating CTE mismatch [×10⁴ K⁻¹] | |
|---|---|---|
| | after 1 h at 1000 °C | after 15 h at 1000°C |
| 30-100 | 2.58 | 3.23 |
| 30-200 | 4.33 | 4.75 |
| 30-300 | 5.29 | 5.68 |
| 30-400 | 5.88 | 6.15 |
| 30-500 | 6.07 | 6.51 |
| 30-600 | 6.33 | 6.63 |
| 30-700 | 6.30 | 6.87 |
| 30-800 | 6.63 | 7.16 |
| 30-900 | 6.99 | 7.49 |
| 30-1000 | 7.20 | 7.88 |

Fig. 4.6.2: Evolution of the CTE of the steel 441 and of the coating system TBC-APP1 after thermal treatment at 1000 °C for 1 and 15 h in air. Measurements up 1000 °C in air, with 10 K min⁻¹. Table: Steel/coating CTE mismatch after thermal treatment at 1000 °C for 1 and 15 h in air up to 1000 °C.

The rather low CTE value of the coating system after 1 h at 1000 °C compared to the estimated value presented in Table 4.2.3 suggests that the thermal expansion, especially of the submicron 3YSZ particles, is somehow accommodated by porosity within the material, reducing the overall expansion of the system.

### 4.6.2 Thermal conductivity

The thermal conductivity of coatings is determined not only by thermal properties of the coating material but also by their microstructural features, such as porosity and cracking pattern. Indeed, not only the absolute values of porosity but also the orientation of cracks and the pore structure (open or closed) have influence on the overall thermal conductivity of the TBC.

After pyrolysis, the coating system is composed mainly of tetragonal-stabilized zirconia and amorphous SiO-based ceramics. Both materials possess low and practically temperature-independent thermal conductivity. While these are typical properties of amorphous glasses, they are not common in crystalline oxides. In the case of tetragonal-stabilized zirconia, this is attributed to the high point defect concentration associated with the substitution of $Zr^{4+}$ ions by $Y^{3+}$ ions, which introduces oxygen vacancies, changing the vibrational modes of zirconia [ClPh2005].

The thermal conductivity of the composite coatings was measured by 3ω method at room temperature and at 500 °C. At room temperature, the average thermal conductivity of the coatings amounts to 0.54 ± 0.14 W m⁻¹ K⁻¹. The increase of the temperature to 500 °C leads to a slightly increase of thermal conductivity up to 0.85 ± 0.08 W m⁻¹ K⁻¹. Schlichting et al. [ScPK2001] have shown that the thermal conductivity of dense 3YSZ decreases gradually from about 3 W m⁻¹ K⁻¹ at room temperature to ~2.7 W m⁻¹ K⁻¹ at 500 °C. Additionally, Cahill [Cahi1990] measured the thermal conductivity of bulk amorphous silica by 3ω method and observed an increase from 1.3 at room temperature to 1.8 W m$^{-1}$ K$^{-1}$ at 500 °C. Thus, due to the predominant presence of tetragonal-stabilized zirconia in the coating system (more than 70 vol% after pyrolysis), a decrease of thermal conductivity was expected with increasing temperature. This suggests that the distribution of the phases and microstructural features of the coatings have significant influence on the thermal conductivity of the coatings.

Indeed, the measured thermal conductivities are significantly lower than those of 3YSZ and amorphous silica. Schlichting et al. [ScPK2001] have shown that, by introducing 10% porosity in 3YSZ, thermal conductivity reduces 0.5 W m$^{-1}$ K$^{-1}$ at all temperatures up to 1000 °C. With a porosity of 18-24%, the thermal conductivity is reduced in a third compared to the dense material. It is important to mention that the porosity of the materials analyzed by the authors was homogeneously distributed and closed, and the samples had no cracks. As previously mentioned, cracks perpendicular to the heat flow are the ideal microstructural feature to reduce thermal conductivity. Due to the presence of a great amount of such cracks in the developed PDC-based top-coat, a significant reduction of the thermal conductivity was expected. Furthermore, Brotzen et al. [BrLB1992] have demonstrated that the thermal conductivity of very thin silica films is considerably lower than that of bulk samples and may reach values far below 1 W m$^{-1}$ K$^{-1}$. As the filler particles are separated by thin silazane-derived SiCNO layers, it is expected that such thin coatings on the particles contribute to a reduction of the overall thermal conductivity. The increase of the thermal conductivity with increasing temperature may be attributed to the high thermal expansion of the passive filler particles. Due to the heterogeneous character of the coatings, the thermal expansion of the passive filler particles may be partially accommodated by pores and cracks, reducing the total porosity, as suggested based on the difference between calculated and measured values of CTE as well (see section 4.6.1). This would result in a reduction of the total porosity and thus of the contribution of the microstructure to the low thermal conductivity.

## 4.7  Durability

### 4.7.1  Thermal shock resistance

The thermal shock resistance of the coatings was investigated by submerging coated samples in liquid nitrogen (T = -196 °C) for 15 min, then immediately transferring the samples to a furnace at 900 °C with another 15 min holding step, constituting one thermal shock cycle. The PDC-based TBC resisted 30 cycles with no signs of spallation. After the 30$^{th}$ cycle, the test was interrupted. Fig. 4.7.1 shows optical microscopy images of a coated sheet before and after the 30 cycles of thermal shock. As one can observe, neither spallation nor cracking could be identified at the surface of the coating.

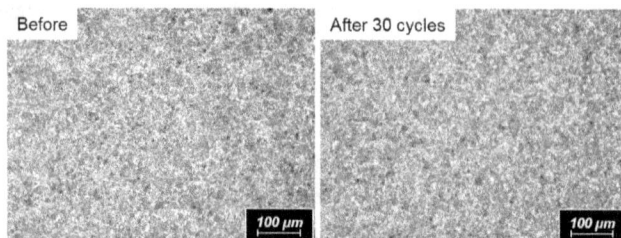

*Fig. 4.7.1: Optical microscopy images of a coating deposited by spraying with formulation TBC-APP1 onto steel 441 followed by pyrolysis at 1000 °C for 1 h, before (left) and after 30 cycles (right) of thermal shock test.*

The excellent thermal shock resistance of the PDC-based TBC is attributed mainly to two characteristics of the system. The first is the strain compliance resulting from porosity and cracks within the coating. These voids minimize thermal stresses by reducing the overall Young's modulus of the coatings. Despite the absence of segmentation cracks (across the thickness), which are the most suitable microstructure to obtain high strain tolerance, the coatings resist severe thermal shocks without failure. The second feature is related to the permanent presence of compressive stress within the coatings, even at high temperatures. In fact, the higher CTE of the substrate relieves part of the compressive stress in the high temperature step of the thermal shock test cycle. Thus, the stresses are lower at high than at low temperatures. In the low temperature stage of the thermal shock cycle, the CTE mismatch between coating and substrate is much lower, so that the generated thermal stresses are not sufficient to cause cracking and spallation of the coatings.

It is, however, important to mention that the tests were carried out with samples subjected to a thermal treatment of only 1 h at 1000 °C during pyrolysis. Hence, the oxidation of the active filler was still not complete. Therefore, the compressive stresses had not yet reached their maximum value, which could reduce the thermal shock resistance. Nevertheless, the thermal shock gradient of almost 1100 °C applied for the tests is higher than any gradient, which might occur in an exhaust system under real conditions, suggesting that the system has a suitable thermal shock resistance for this application.

### 4.7.2 Long-term temperature and oxidation resistance

The oxide scale of the ferritic steel 441 is formed by two layers [CWAT2007, RGWD2008, SCGP2009, IMCP2012]. Directly on the surface of the substrate, a chromium-rich oxide layer grows, above which a second layer composed of Cr-Mn spinel forms. After prolonged oxidation, a silicon oxide layer forms between the Cr-rich oxide scale and the substrate. Moreover, segregation of Ti and Nb oxides occurs, at first near the interface metal/oxide scale. With the progress of the oxidation, this oxides diffuse through the Cr-rich layer and migrate to the surface [GCTW2008, JaCS2010, GrFS2014]. Furthermore, according to Jablonski et al. [JaCS2010] and Grolig et al. [GrFS2014], Laves phases ($Fe_2Nb$) form within the metal, which also capture silicon atoms, competing with the formation of the silica scale.

The resistance of coated parts to prolonged exposure to high temperatures in oxidative environment was investigated by subjecting coated and uncoated samples to a temperature of 900 °C in air for 10, 50 and 100 h. The cross-section of the samples was investigated by SEM and EDS in order to evaluate the stability of the system.

Fig. 4.7.2 presents SEM images with EDS element mappings of uncoated stainless steel 441 after oxidation at 900 °C in air for 10 h. The SEM image reveals the growth of an inhomogeneous oxide scale (TGO) with large oxide crystals at the surface. After 10 h, the TGO has a thickness of ~2 μm. According to Chandra-Ambhorn and colleagues [CWAT2007], the oxidation of steel grade 441 follows a parabolic rate law and is limited by the rate of chromium diffusion within the substrate and/or through the TGO. The EDS mappings confirmed the expected formation of a double-layered TGO, with a chromium-rich layer directly on the substrate and a manganese-rich layer onto the Cr-rich one, whereas iron was not detected within the oxide scale. Titanium was detected at or near the surface of the TGO, in agreement with previous studies [GCTW2008, JaCS2010, GrFS2014]. Si mapping (not shown) revealed some silicon enrichment at the interface metal/TGO, as previously reported elsewhere [JaCS2010, IMCP2012].

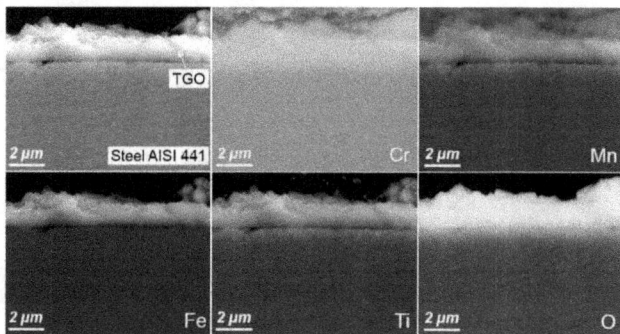

Fig. 4.7.2: SEM micrographs with EDS element mappings of the cross-section of an uncoated stainless steel 441 sample after oxidation at 900 °C in air for 10 h.

Samples coated with the PDC-based TBC with and without the PHPS bond-coat were likewise investigated (Fig. 4.7.3). After 10 h at 900 °C in air, the coated sample without bond-coat developed a TGO similar to that of uncoated samples, with the Cr-Mn double-layer configuration. Titanium was also detected close to the interface steel/TGO and near the interface TGO/top-coat, which is detrimental to the adhesion of the layers [PaGJ2002]. Despite the presence of silicon in the top-coat, no diffusion of this element toward the substrate was detected. Moreover, the thickness of the TGO was also comparable to that of uncoated samples (~2 μm), evidencing that the top-coat offers no significant oxidation protection. This was expected owing to the high oxygen permeability of 3YSZ and the low amount of silazane in the formulation. Silicon segregation at the interface metal/TGO was not observed.

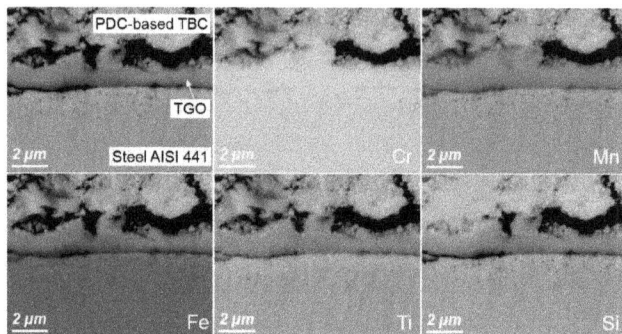

Fig. 4.7.3: SEM micrographs with EDS element mappings of the cross-section of a stainless steel 441 sample coated with formulation TBC-APP1 (without PHPS bond-coat) after oxidation at 900 °C in air for 10 h.

However, a different morphology was observed on the cross-section of samples coated both with bond-coat and top-coat (Fig. 4.7.4).

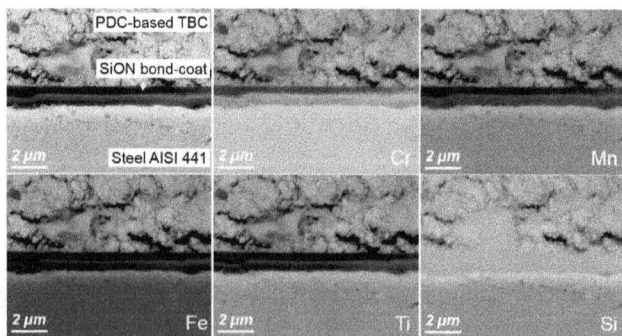

Fig. 4.7.4: SEM micrographs with EDS element mappings of the cross-section of a stainless steel 441 sample coated with PHPS and formulation TBC-APP1 after oxidation at 900 °C in air for 10 h.

Diffusion of elements from the substrate into the bond-coat and toward the top-coat, as observed already during pyrolysis for 1 h at 1000 °C (see Fig. 4.4.1d), led to the formation of a metal-containing SiON bond-coat (comparison between Cr, Mn and Si mappings). Interestingly, this diffusion occurred uniformly, whereas the protrusions observed after pyrolysis are not present anymore. In this case, the typical distribution of

Cr and Mn is not evident, and the diffusion layer seems to be composed of homogeneously distributed Cr and Mn. Although oxygen could also be detected within the diffusion layer (mapping not shown), it is not possible to affirm that the metals are oxidized, since oxygen may be associated only to the SiON bond-coat. Titanium was not detected within the bond-coat, although titanium segregation has occurred within the substrate near the interface with the bond-coat. The thickness of the diffusion zone amounts to ~0.5 μm. This proves that the PHPS-derived layer is capable of delaying the growth of the oxide scale significantly, due to the low diffusion rates of oxygen and of the metallic elements through the SiON bond-coat, although it is not capable of avoiding the latter completely.

The second oxidation test series was performed with 50 h at 900 °C in air. The TGO of uncoated steel samples have similar distribution of elements (EDS mappings not shown) like the ones after 10 h of oxidation at 900 °C. However, a further growth of the layer reduced the mechanical stability and adhesion of the TGO, whereas cohesion and adhesion failure mechanisms occurred in large areas during metallographic procedures (Fig. 4.7.5a). Hence, despite the apparent stabilization of the thickness at ~2 μm (Fig. 4.7.5b), it is possible that a part of the layer has spalled off the substrate during the oxidation tests.

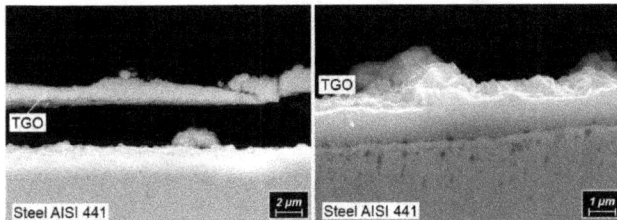

*Fig. 4.7.5: SEM images of the cross-section of an uncoated stainless steel 441 sheet after oxidation at 900 °C in air for 50 h.*

After 50 h at 900 °C in air, samples coated only with the top-coat (Fig. 4.7.6) have similar elemental distribution as observed after 10 h of oxidation. The thickness of the TGO exceeds 2 μm, confirming that the TGO of the uncoated samples was indeed damaged during the oxidation test or during sample preparation for SEM analysis. The most interesting feature of samples coated only with formulation TBC-APP1 after 50 h at 900 °C in air is the drastically changed metal/TGO interface, which has been clearly deteriorated. Similar changes of interface roughness were observed by Wang et al. [WGMF2013] after long-term oxidation of a nickel-based superalloy coated with siloxane/ZrSi$_2$ mixtures. At this interface, an enrichment of silicon, chromium and titanium was detected, which is in accordance with the findings of Jablonski [JaCS2010].

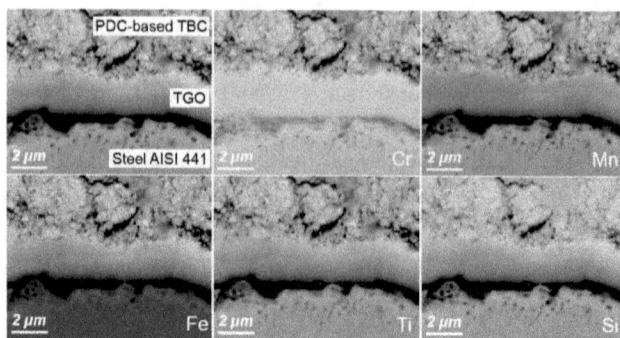

Fig. 4.7.6: SEM micrographs with EDS element mappings of the cross-section of a stainless steel 441 sample coated with formulation TBC-APP1 (without PHPS bond-coat) after oxidation at 900 °C in air for 50 h.

In contrast, samples coated with a PHPS bond-coat and the TBC show no sign of spallation after 50 h of oxidation at 900 °C, and the interface between substrate and bond-coat/TGO remains homogenous, with no significant roughness increase (Fig. 4.7.7). The most interesting change compared to samples oxidized for 10 h at 900 °C in air is the formation of the oxide scale above the SiON layer, most likely due to diffusion of metals across the whole thickness of the SiON bond-coat. The oxide scale has now the characteristic double-layered structure and a thickness of about 1 μm. Chromium can also be detected within the SiON bond-coat, whereas Mn is almost exclusively detected within the TGO. Titanium is now also detectable within the SiON bond-coat, although in low quantity. Interestingly, the thickness of the SiON layer decreases significantly, which may be related to the diffusion of silicon toward the substrate.

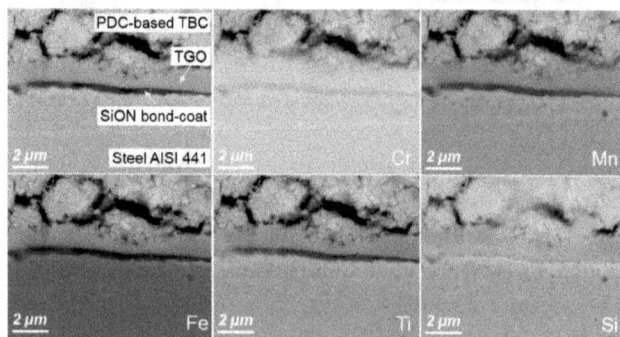

Fig. 4.7.7: SEM micrographs with EDS element mappings of the cross-section of a stainless steel 441 sample coated with PHPS and formulation TBC-APP1 after oxidation at 900 °C in air for 50 h.

After 100 h of oxidation, the uncoated substrate was not analyzed by SEM/EDS due to the fragility of the TGO. Samples with TBC but without bond-coat, despite the apparent stability after the oxidation test, did not withstand the metallographic procedure and detached completely from the substrate (Fig. 4.7.8a). After 100 h, the TGO is as thick as 5 μm. The spallation at the interface substrate/TGO evidences that the adhesion of the top-coat to the TGO is higher than the adhesion of the TGO to the substrate, which may be associated with the formation of the silicon oxide-based interlayer underneath the Cr-rich oxide scale.

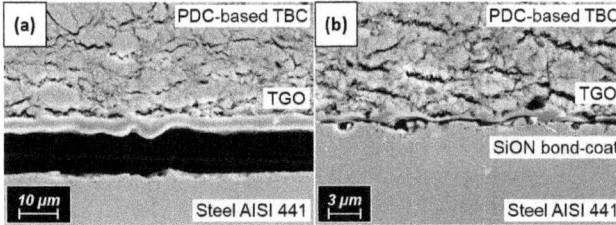

Fig. 4.7.8: SEM micrographs of the cross-section of stainless steel 441 coated with TBC-APP1 after oxidation at 900 °C in air for 100 h: (a) without PHPS bond-coat and (b) with PHPS bond-coat.

Samples coated with PHPS and the top-coat show no sign of spallation after 100 h at 900 °C in air (Fig. 4.7.8b). However, a significant morphology change at the interface coating/substrate occurs – the roughness of the interface increases, and pores are formed.

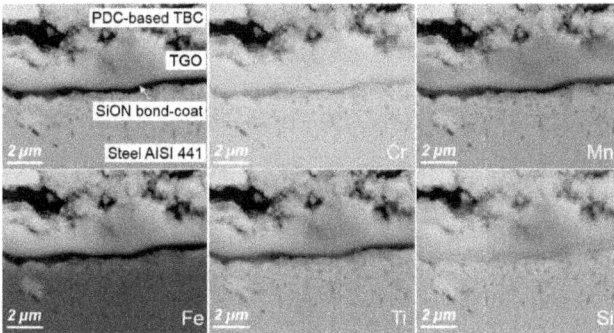

Fig. 4.7.9: SEM and EDS images of the cross-section of a stainless steel 441 sample coated with PHPS and formulation TBC-APP1 after oxidation at 900 °C in air for 100 h.

Despite this change at the interface, no adhesion failure takes place. This may be attributed to the presence of the SiON layer, which acts like an adhesive between substrate and TGO, as evidenced by the EDS mappings (Fig. 4.7.9). The EDS mappings also revealed the presence of titanium near the interface TGO/top-coat. The thickness of the oxide-scale is between 1 and 2.5 μm and is characterized by low homogeneity. The white spots within the substrate visible in the SEM images identify the Laves phases. As visible in the EDS mappings, silicon and titanium are captured by this phase, reducing the diffusion of these elements toward surface.

As the investigations showed, the coatings can overcome stresses caused by long-term exposure to high temperatures in an oxidative environment and thermal shocks. However, the limiting factor regarding durability seems to be the resistance of the system against oxidation. The drastic changes at the interface caused by an inhomogeneous growth of the TGO and by diffusion of certain elements from the substrate through the bond-coat most likely change both adhesion and distribution of stresses in the system, which may result in coating failure during application.

## 4.8   Coating on the inside of pipes

### 4.8.1   Deposition of coatings

To enable the application of the PDC-based TBCs in automotive exhaust systems, the coating of the inner face of pipes must be possible. Due to the small diameter of the exhaust pipes (53 mm), deposition of coatings onto the inner face is a great challenge. Pipes with length of 500 mm were firstly coated by dip coating with PHPS, followed by pyrolysis at 500 °C for 1 h. Then, the top-coat was deposited using a home-made coating apparatus, as described in section 3.3, with air as carrier gas. A second pyrolysis was performed at 1000 °C for 1 h in a chamber furnace. Fig. 4.8.1a presents a digital photograph of the outside of uncoated and coated pipes after pyrolysis at 1000 °C. Once again, it becomes clear that the PHPS layer offers an excellent oxidation protection to the ferritic stainless steel AISI 441, which not only improves the appearance of the exhaust pipes but also increase their lifespan. Fig. 4.8.1b shows a digital photograph of the inside of a coated pipe. The surface of the top-coat has a smooth and homogeneous appearance and no spallation was detected, not even on the pipe seam. The reproducibility of the processing, however, still must be improved. Changes in the suspension flow occur from one deposition to the other, especially when several pipes are coated in sequence. This may occur due to accumulation of the coating suspension in the feeding hoses and spray lance, which alter the flow of the suspension. Furthermore, the use of air as carrier gas may lead to cross-linking of the silazane, blocking partially the spray nozzle. Despite these difficulties, the deposition of PDC-based TBCs onto the inner face of pipes was proven to be feasible.

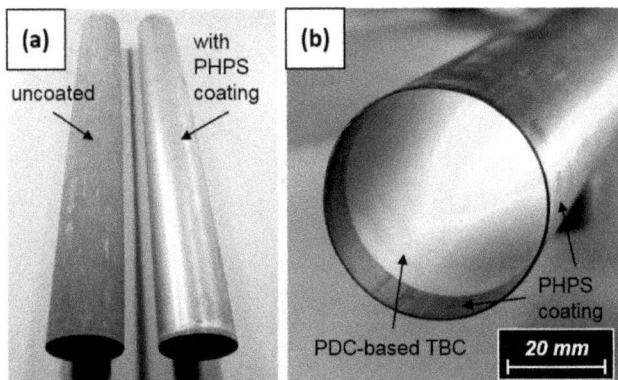

Fig. 4.8.1: Digital images of pipes after treatment at 1000 °C for 1 h: (a) outside of an uncoated and of a coated pipe, L = 500 mm, $\varnothing_i$ = 53 mm; (b) inside of a coated pipe.

### 4.8.2   Investigation of the insulating effect

The insulating effect of the PDC-based TBCs deposited onto the inner face of pipes was investigated by measuring the temperature at the outer face of coated and uncoated pipes, while hot air flew upstream inside the pipes. Fig. 4.8.2 presents the data recorded by the data-logger. The measurements revealed that, despite the low coating thickness compared to conventional TBCs, the PDC-based TBCs have a significant insulating effect, evidenced by the difference in the outer face temperature of the coated and uncoated pipes. After about 25 s, the temperature at the outer face of coated and uncoated pipes becomes different. After 120 s, which is similar to the duration of the light-off phase of the catalytic converter, the temperature difference was about 5 K. The low thermal conductivity of the TBC reduces the heat transfer rate from the hot gas stream to the outer surface of the metal pipe. At the same time, the metal pipes exchange heat with the environment, resulting in a lower temperature at the surface of the coated pipes when compared to uncoated ones. It is important to mention that during the tests, turbulence at the gas outlet may result in the heating of the pipe from the outside as well. This heat is then transported by conduction to the measurement spot, increasing the measured wall temperature. Therefore, it is expected that the real temperature difference would be significantly larger in a completely isolated system, where heat may only be transferred from the inside of the pipes by conduction across the coating.

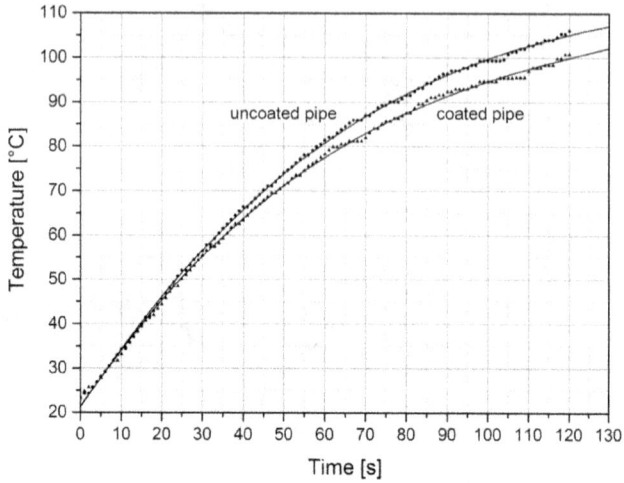

Fig. 4.8.2: Evolution of the outer face temperature of coated and uncoated pipes during the flow of hot air on the inside of the pipes.

## 5 SUMMARY

Thermal barrier coatings (TBCs) are systems developed to reduce the heat transfer from a hot medium to the coated parts. This causes a reduction of the part's temperature and/or a lower heat loss from the hot medium. Such coatings have been deposited by methods like CVD, PVD and thermal spraying techniques. These techniques have disadvantages such as relatively expensive and complex processing, as well as a limited flexibility regarding part geometry. An alternative approach is the polymer-derived ceramics (PDC) route. In this technique, precursors, usually silicon-based polymers, are transformed into ceramic materials by a thermal treatment. The PDC technology is a versatile method to prepare ceramic coatings, with the advantages of low processing costs, easy processability, flexibility regarding substrate geometry, and relatively easy automation. However, the greatest drawback of the technique is the high shrinkage of precursors during conversion into ceramics, which cause stresses in the coatings, limiting their maximum thickness. Moreover, the low CTE of silicon-based ceramics makes the development of coatings for metallic substrates for high temperature applications challenging. Due to the CTE mismatch between coating and substrate, additional stresses arise in the system, especially during large temperature variations, which can lead to coating failure.

The objective of the present work was the development of a TBC for automotive exhaust systems – thus requiring stability up to ~950 °C – by PDC processing. These coatings should be able to reduce the heat transfer from the exhaust gases to the metal pipes, resulting in a faster heating-up of the catalytic converter and, thus, more efficient conversion of hazardous gases during the first seconds of engine operation. An efficient insulating effect depends on a low thermal conductivity and a high thickness of the coatings. Thus, the objective of the work was to obtain a coating system with thermal conductivity similar to that of conventional TBCs with thickness as large as possible. Due to the heterogeneity of the system, which combines ceramics, metals and polymers, the development of thick PDC coatings with tailored properties and high temperature stability is difficult. The preparation of such coating systems becomes possible only by the selection of suitable materials, a fine adjustment of composition, and the optimization of deposition process and thermal treatments.

Aiming at the application in automotive exhaust systems, ferritic stainless steel grade 441 was selected as substrate. This steel has been applied by the automotive industry to produce exhaust pipes due it its suitable temperature and oxidation stability, good weldability and conformability, and reduced price compared to other stainless steel grades and high temperature alloys. Another important advantage of this ferritic steel grade is the lower linear coefficient of thermal expansion (CTE). While CTEs of other grades used in the production of exhaust systems reach values around $20\times10^{-6}$ $K^{-1}$ up to 1000 °C, the CTE of the steel grade 441 remains below $14\times10^{-6}$ $K^{-1}$. Upon oxidation, this steel forms a double-layered oxide scale, also called thermally grown oxide layer (TGO). The layer immediately on the substrate is composed mainly of chromium oxide, whereas the external layer is composed of Cr-Mn spinel.

To improve the compatibility between TBC and substrate, a thin bond-coat was deposited and pyrolyzed before deposition of the insulating top-coat. This thin layer was composed of silazane PHPS and was deposited by dip coating, followed by pyrolysis in air at 500 °C for 1 h. After pyrolysis, a SiON ceramic coating with high surface energy is obtained, which improves the oxidation resistance of the system. The high surface energy ensures a good wetting with the second coating suspension.

The main component of the developed TBC top-coats is yttria-stabilized zirconia (YSZ), which acts as a passive filler in the system. The function of YSZ is to reduce the thermal conductivity and to increase the CTE of the coating. Another component of the PDC-based TBC system is zirconium disilicide ($ZrSi_2$). $ZrSi_2$ oxidizes in air at temperatures above 450 °C to form zirconia and silica, which causes a volume expansion. Thus, $ZrSi_2$ acts as an active filler in the system, compensating the shrinkage of the precursor. Together, active and passive fillers minimize the stresses caused by precursor shrinkage during pyrolysis, enabling preparation of thicker coatings. Durazane 1800 was selected as precursor for the preparation of the top-coats. This organosilazane acts as a temperature-resistant binder for the filler particles. Moreover, it enables the deposition of the coatings in liquid phase under air. Coatings were deposited by doctor blade and spray coating techniques onto ferritic steel 441 substrates. To convert the preceramic polymer into a ceramic material, generating purely ceramic top-coats, pyrolysis was conducted in air at 1000 °C for 1 h.

Coating formulations with different compositions were applied to study the influence of each filler. These investigations have shown that the system is very sensitive to composition, whereas not only the ratio of precursor to active filler is decisive, but also the amount of passive filler. The best coatings were obtained with 64 vol% YSZ, 27 vol% silazane and 9 vol% $ZrSi_2$ in the starting formulation (disregarding solvent and dispersant amounts). Coatings with maximum thickness of ~50 μm were prepared by single deposition and pyrolysis procedures. This coating system was characterized regarding microstructure, adhesion, thermal expansion and conductivity, and durability.

The behavior of the coating system during pyrolysis was investigated by conventional dilatometry on dense monolithic samples of coating material and by thermo-optical dilatometry on coated steel sheets. These investigations revealed that the active filler not only compensates the shrinkage of the silazane, but also causes an expansion of the coatings, which is larger than the thermal expansion of the steel substrate. This results in the generation of compressive stresses within the coatings. However, according to calculations based on mass and density changes during pyrolysis, and on the CTE of the individual components, the coating system should not undergo expansion during pyrolysis. This is attributed to the different temperature ranges where the expansion of the active filler and the cross-linking of the silazane take place. Durazane 1800 becomes a thermoset polymer at temperatures as low as 130 °C. However, the oxidation of the active filler begins only at ~450 °C. Thus, the expansion of active filler particles cannot be accommodated in the surrounding SiCNO/YSZ coating matrix by viscous flow and the expansion causes the formation of cracks and an overall expansion of the material.

Dilatometry investigations on uncoated steel samples have also shown that the thermal expansion of the steel increases from $10.1 \times 10^{-6}$ $K^{-1}$ in the range of 30-200 °C to $13.3 \times 10^{-6}$ $K^{-1}$ in the range of 30-1000 °C. The CTE of pyrolyzed monolithic samples of coating material, on the other hand, is virtually constant ($5-6 \times 10^{-6}$ $K^{-1}$) in the same interval, causing an increase of CTE mismatch between substrate and coating with increasing temperature. Moreover, the lower CTE of the coating material measured by dilatometry compared to calculated values based on the CTE and the amount of each component in the coating system ($9.1 \times 10^{-6}$ $K^{-1}$) indicates that part of the thermal expansion is accommodated by cracks and pores.

Since coating expansion caused by the active filler is irreversible, and the CTE of the metal is larger than that of the coating, the metal retracts more than the coating upon cooling after pyrolysis. This increases the compressive stresses in the coatings, explaining the absence of segmentation cracks and the presence of diagonal cracks in the microstructure of coatings after pyrolysis. The obtained microstructure is composed mainly of cracks parallel to the substrate. While the expansion of the coating parallel to the substrate is constrained by the adhesion to the metal sheet, the expansion normal to the substrate (across the thickness) is not constrained. This expansion then generates tensile stresses, leading to formation of cracks parallel to the substrate. A longer exposure of the coatings to high temperatures in air leads to further expansion of the active filler, increasing the compressive stresses. After about 10 h at 1000 °C, the expansion finally ceases.

Adhesion investigations revealed that the coating system has strong adhesion and cohesion, attributed to covalent bonds between the silazane and the substrate and filler particles. Cross-cut tape tests were performed, and the coatings were classified as Gt-1 (removal of up to 5% of the tested area) after visual inspection. However, an evaluation by optical microscopy revealed that cohesion failure was the dominant failure mechanism, which resulted in damage of about 51% of the tested area during the test. Pull-off tests resulted in an average tensile adhesion of 20.9 MPa, which is in the range of values of conventional TBCs. The occurrence of both adhesion and cohesion failure mechanisms was observed.

Thermal conductivity measurements by $3\omega$ method revealed the development of coatings with very low thermal conductivity. The combination of low thermal conductivity of the components with the excellent microstructure of the coatings resulted in thermal conductivities of $0.54 \pm 0.14$ W m$^{-1}$ K$^{-1}$ at room temperature and $0.85 \pm 0.08$ W m$^{-1}$ K$^{-1}$ at 500 °C, which are below the average thermal conductivity values of conventional TBCs ($0.8$-$1.7$ W m$^{-1}$ K$^{-1}$).

Durability investigations on PDC-based TBCs revealed a high thermal shock resistance and long-term oxidation stability. The coatings resisted 30 cycles of thermal shock with a temperature difference of ~1100 °C with no signs of spallation. Moreover, the PHPS bond-coat was proven to be crucial to the oxidation stability of the system. After 100 h at 900 °C in air, top-coats deposited onto uncoated substrates are barely adherent, while the ones deposited onto substrates with a PHPS-derived SiON bond-coat are stable. Furthermore, the thickness of the oxide scale is reduced from ~5 μm to less than 2.5 μm owing to the PHPS bond-coat. Initially Cr and Mn and, after longer time also Ti, are able to diffuse through the SiON layer, forming an oxide scale between bond-coat and top-coat. The SiON coating then acts as an adhesive layer, preventing spallation of the TGO. Despite the reduced oxidation obtained with the bond-coat, the interface with the substrate changes significantly after 100 h at 900 °C, owing to an inhomogeneous diffusion of metallic elements through the bond-coat.

To enable deposition of the PDC-based TBCs on the inside of pipes, a spray equipment was developed and coatings were deposited inside of exhaust pipes with length as large as 500 mm and diameter as small as 53 mm. Investigations on the thermal insulation proved that, despite the low thickness in comparison to conventional TBCs, the PDC-based TBC is able to reduce the heat transfer from a hot air flowing on the inside of the metal pipes to the pipe walls. Owing to the very low thermal conductivity of the coatings, the wall temperature at the outside of coated pipes was ~5 K lower when compared to uncoated pipes.

# 6 Outlook

The developed coating system has great potential for application at temperatures below 1000 °C. However, some characteristics of the system can still be optimized to improve performance and increase the lifespan of the coatings. The most critical aspects of the developed coating system seem to be the limited coating thickness and the oxidation of the substrate.

To improve coating thickness, a better control of the stresses arising during pyrolysis and application is necessary. The first strategy to reduce these stresses is based on the introduction of a sacrificial filler. The elimination of the sacrificial filler during pyrolysis generates additional porosity, which reduces the overall Young's modulus of the coatings, consequently increasing the strain compliance of the system. A fine adjustment of the volume fractions of sacrificial and active filler may result in coatings with larger coating thickness and reduced thermal conductivity.

The second strategy is to adjust the expansion of the active filler. $ZrSi_2$ was selected based on its large volume expansion, in order to obtain a high volume increase with a small amount of active filler. Thus, high amounts of 3YSZ, responsible for thermal properties of the system, could be used. However, the oxidation of $ZrSi_2$ begins only at ~450 °C, while the shrinkage of the silazane starts at temperatures around 120 °C. As was observed by dilatometry, the coating system initially undergoes a shrinkage, followed by an expansion during the high temperature phase of pyrolysis. Moreover, the pyrolysis duration and temperature are not sufficient to complete the expansion of the filler, whereas it continues to expand during the following hours at high temperatures, increasing the compressive stresses in the coatings. Despite the significantly lower theoretical volume expansion compared to $ZrSi_2$, $ZrC$ oxidizes faster, with onset temperature around 250 °C. The mass increase reaches its maximum value of approx. 20% at ~600 °C, which then diminishes slightly up to 1000 °C due to the elimination of carbon. Thus, by substituting partially $ZrSi_2$ for $ZrC$, the expansion of the fillers may be adjusted to match with the shrinkage of the silazane during the whole pyrolysis procedure. $ZrC$ may be added in sufficient amounts to compensate precursor shrinkage in the low temperature range while $ZrSi_2$ is responsible for the high temperature range. By reducing the volume changes during pyrolysis, also the stresses are reduced, possibly enabling the preparation of thicker coatings. The elimination of carbon from $ZrC$ may also induce the formation of additional transient and/or permanent porosity, which could increase the strain compliance of the system and reduce the dimensional changes.

A third approach may be suggested, which consists in the combination of the first and second approaches. However, due to the elevated number of components, the development of this system may be a great challenge, requiring a large experimental effort to optimize the behavior of the coatings during pyrolysis.

The second main issue is related to the oxidation of the substrate. As shown, the PHPS-derived bond-coat is not able to prevent the diffusion of metallic elements from the substrate toward the surface, whereas a TGO layer is formed between bond-coat and top-coat. This TGO layer is responsible for a reduction in the adhesion of the coating system and for coating failure upon long-term exposure to high temperatures in air. Progress may be made simply by improving the surface quality of the steel. As shown, the roughness of the substrate causes localized failure of the PHPS layer, which facilitates the growth of the TGO layer. The development of a thicker PHPS bond-coat might also be an interesting approach. This would require, however, multiple deposition and pyrolysis procedures, increasing the processing time and cost. Moreover, even with multiple depositions, the coating thickness is still limited to a few microns due to the CTE mismatch. A multilayered system composed of different layers may also be considered. The layer applied onto the substrate should be responsible for blocking the diffusion of metallic elements, while the PHPS bond-coat, applied onto the primary layer, blocks the inward diffusion of oxygen. However, the same issues related to the processing of multiple PHPS layers persist. Thus, the development of a thicker, single-layered bond-coat with high CTE, capable of preventing not only the inwards diffusion of oxygen but also the diffusion of the metallic elements seems to be a more definitive solution to the durability issue. The development of a particle-filled PDC bond-coat might be a promising strategy. By means of glass fillers, thick and dense coatings with high thermal expansion may be obtained. So far, such coatings were developed only for application temperatures below 800 °C. Thus, a significant research effort is still required to improve the properties of the bond-coat without excessively increasing process complexity.

Another improvement, required for the deposition of the coatings inside of pipes is the development of an optimized spray equipment. An improved nozzle configuration may enable a better control of the deposition, increasing the reproducibility. The use of inert gas as carrier medium may improve the process by preventing premature cross-linking of the precursor. Moreover, the development of a compact spray equipment, constituted only by a small nozzle unit with an auto-centering mechanism, connected directly to a flexible feeding hose may enable even the deposition of coatings inside of curved pipes.

# 7 ZUSAMMENFASSUNG

Wärmedämmschichten (TBC, thermal barrier coating) sind Beschichtungssysteme, die für eine Reduzierung der Wärmeübertragung von einem Heißgas an ein vorwiegend metallisches Substrat sorgen, was zu einer Verringerung der Substrattemperatur und/oder des Wärmeverlusts führt. Solche Beschichtungen werden meistens mittels chemischer oder physikalischer Gasphasenabscheidung (CVD bzw. PVD, chemical bzw. physical vapor deposition) sowie über das thermische Spritzen hergestellt, die neben den relativ hohen Investitions- und Betriebskosten Einschränkungen hinsichtlich der beschichtbaren Substratgeometrie aufweisen. Eine alternative Methode keramische Schichten zu erzeugen ist die sogenannte „polymerabgeleitete Keramik"-Technologie (PDC, polymer-derived ceramic). Dabei werden Precursoren, meistens siliziumbasierte Polymere, durch eine thermische Behandlung (Pyrolyse) in keramische Werkstoffe umgewandelt. Die PDC-Technologie wurde bereits zur Herstellung unterschiedlichster Beschichtungen eingesetzt. Vorteilhaft sind hierbei vor allem die relativ niedrigen Investitions- und Betriebskosten, die einfache Handhabung, eine erhöhte Flexibilität hinsichtlich der Substratgeometrie und die einfache Automatisierung des Beschichtungsprozesses.

Allerdings weist die PDC-Technologie auch einen wichtigen Nachteil auf. Aufgrund des Dichtanstiegs und der Massenverluste während der Pyrolyse, schrumpft das Material dermaßen, dass Spannungen innerhalb der Beschichtungen entstehen, die die maximale Schichtdicke begrenzen. Darüber hinaus weisen siliziumbasierte Keramiken meistens einen geringeren thermischen Ausdehnungskoeffizienten (TAK) im Vergleich zu den metallischen Substraten auf, was beim Temperaturwechsel zu zusätzlichen Spannungen führt. Daher ist die Entwicklung dicker Schichten eine große Herausforderung.

Das Ziel der Doktorarbeit war die Entwicklung einer Wärmedämmschicht auf Basis der PDC-Technologie für die Anwendung in automotiven Auspuffsystemen, wobei Temperaturen von bis zu ~950 °C auftreten können. Dabei sollen die Schichten die Wärmeübertragung von den Heißgasen zu den metallischen Rohren reduzieren, wodurch eine schnellere Aufheizung des Katalysators und damit eine effizientere Umsetzung der Schadstoffe erfolgen soll. Die Isolationswirkung der Beschichtungen hängt insbesondere von einer niedrigen Wärmeleitfähigkeit und hohen Schichtdicken ab. Daher sollen möglichst dicke Schichten entwickelt werden, die eine mit konventionellen Wärmedämmschichten vergleichbare Wärmeleitfähigkeit aufweisen. Aufgrund der Heterogenität des Beschichtungssystems, das polymere, keramische und metallische

Werkstoffe kombiniert, ist die Entwicklung von Beschichtungen, die gleichzeitig eine hohe Temperaturstabilität und funktionelle Eigenschaften aufweisen, äußerst schwierig.

Aufgrund der speziellen Anwendung in automotiven Auspuffsystemen wurde der ferritische Edelstahl 1.4509 (AISI 441), der für die serienmäßige Herstellung von Auspuffsystemen eingesetzt wird, als Substratmaterial ausgewählt. Dieser Stahl weist, neben einem geringeren Preis im Vergleich zu anderen Edelstählen für hohe Anwendungstemperaturen, eine gute Schweiß- und Umformbarkeit sowie eine hohe Temperatur- und Oxidationsstabilität auf. Einen zusätzlichen Vorteil des Stahls 1.4509 ist der geringe TAK. Während der TAK anderer für die Herstellung von Auspuffsystemen eingesetzter Stähle Werte von $20 \times 10^{-6}$ $K^{-1}$ erreicht, beträgt er für den Stahl 1.4509 weniger als $14 \times 10^{-6}$ $K^{-1}$ bis zu 1000 °C. Während der Oxidation des Stahls bildet sich eine zweischichtige, thermisch gewachsene Oxidschicht (TGO), bestehend hauptsächlich aus Chromoxid und die darauf liegende Schicht aus Cr-Mn-Spinellen.

Um die Kompatibilität der PDC-basierten Wärmedämmschichten mit dem Substrat zu verbessern, wird vor ihrer Applizierung eine dünne Zwischenschicht (sog. Bond-Coat) aufgetragen und pyrolysiert. Diese dünne Schicht besteht aus dem Silazan PHPS und wird mittels Tauchbeschichtung aufgetragen. Die Pyrolyse dieser Schicht erfolgt bei 500 °C für 1 h, um das Silazan in eine SiON Keramikbeschichtung mit hoher Oberflächenenergie umzuwandeln, die für eine gute Benetzbarkeit der Oberfläche sorgt. Darüber hinaus bietet die SiON-Schicht einen hervorragenden Oxidationsschutz bereits während der Pyrolyse der Wärmedämmschicht.

Der Hauptbestandteil des Beschichtungssystems ist Yttriumoxid-stabilisiertes Zirkonoxid (YSZ), das als passiver Füllstoff dient. Die Aufgaben des YSZ sind, die Wärmeleitfähigkeit der Beschichtung zu reduzieren und den TAK zu erhöhen. Eine andere Komponente des Beschichtungssystems ist Zirkondisilizid ($ZrSi_2$), das ab einer Temperatur von 450 °C oxidiert und zur Bildung von Zirkonoxid und Silika führt, was mit einer Volumenzunahme verbunden ist. Daher ist $ZrSi_2$ ein aktiver Füllstoff, der der Schrumpfung des Precursors entgegenwirkt. Zusammen minimieren passive und aktive Füllstoffe die gesamte Schrumpfung der Beschichtung und verringern dadurch die entstehenden Spannungen, was die Herstellung von dickeren Schichten ermöglicht. Als Precursor bei der Herstellung der PDC-Beschichtungen diente das Silazan Durazane 1800, das das Auftragen der Beschichtungen an Luft durch Nassverfahren ermöglicht und als temperaturbeständiges Bindemittel fungiert. Die Beschichtungen wurden mittels Rakel oder Aufsprühen auf die Stahlsubstrate appliziert. Um den Silazanprecursor in eine Keramik umzuwandeln, wurde anschließend eine Pyrolyse bei 1000 °C für 1 h an Luft durchgeführt.

Beschichtungsformulierungen mit unterschiedlichen Zusammensetzungen wurden entwickelt, um den Einfluss der jeweiligen Komponente im System zu untersuchen. Es zeigte sich, dass das System eine hohe Sensibilität bezüglich der Zusammensetzung aufweist, wobei nicht nur das Verhältnis zwischen den Volumenanteilen von aktivem Füllstoff und Precursor, sondern auch der Anteil an passiven Füllstoffen eine Rolle hinsichtlich der Stabilität der Beschichtungen spielt. Die besten Beschichtungen wurden mit einer Formulierung bestehend aus 64 Vol.% 3YSZ, 27 Vol.% Silazan und 9 Vol.% $ZrSi_2$ erhalten. Beschichtungen mit einer Schichtdicke von ca. 50 µm konnten so durch eine einzige Applizierung und Pyrolyse hergestellt werden. Diese Beschichtungen wurden hinsichtlich ihrer Mikrostruktur, Haftfestigkeit, thermischen Ausdehnung, Wärmeleitfähigkeit sowie Temperaturwechsel- und Oxidationsbeständigkeit untersucht.

Um das Verhalten des Beschichtungssystems während der Pyrolyse zu untersuchen, wurden konventionelle und thermo-optische Dilatometrieanalysen an monolithischen Proben aus dem Beschichtungsmaterial bzw. an beschichteten Stahlsubstraten durchgeführt. Dabei zeigte sich, dass der aktive Füllstoff nicht nur die Schrumpfung des Precursors während der Pyrolyse kompensiert, sondern auch zu einer Ausdehnung der monolithischen Proben bzw. der Beschichtungen führt, die größer als die thermische Ausdehnung des Stahls ist. Daraus ergeben sich Druckspannungen innerhalb der Beschichtungen. Auf Basis von Berechnungen, die die Massen- und Dichteänderungen sowie die thermischen Ausdehnungskoeffizienten der einzelnen Komponenten berücksichtigten, wurde allerdings eine Schrumpfung des Beschichtungsmaterials erwartet. Dieser Unterschied zwischen dem kalkulierten und gemessenen Schrumpfungs-/Dehnungsverhalten ergibt sich aus unterschiedlichen Temperaturen, bei denen die Umwandlung der Komponenten stattfindet. Durazane 1800 vernetzt bereits ab Temperaturen von ca. 130 °C und ist bei 700 °C vollständig in eine Keramik umgewandelt, während die Oxidation des $ZrSi_2$ erst ab 450 °C beginnt. Daher kann sich die PDC/YSZ-Matrix nicht durch einen viskosen Fluss an die Ausdehnung des aktiven Füllstoffes anpassen, wodurch Risse und Poren entstehen, was zu einer Expansion des gesamten Beschichtungssystems führt.

Die Dilatometrieuntersuchungen an unbeschichteten Substraten zeigten, dass der TAK des Stahls 1.4509 von $10{,}1 \times 10^{-6}$ $K^{-1}$ bis 200 °C auf $13{,}3 \times 10^{-6}$ $K^{-1}$ bis 1000 °C steigt. Im Gegensatz dazu ist der TAK von monolithischen Proben aus dem Beschichtungsmaterial nach der Pyrolyse praktisch konstant ($5\text{-}6 \times 10^{-6}$ $K^{-1}$), wodurch sich der TAK-Unterschied und damit auch die Spannungen mit zunehmender Temperatur vergrößern. Allerdings sind die gemessenen TAK-Werte des Beschichtungsmaterials wesentlich geringer, als der, auf der Basis der Zusammensetzung und des TAK der einzelnen Komponenten, berechnete Wert von $9,1\times10^{-6}$ $K^{-1}$. Dies weist darauf hin, dass die thermische Ausdehnung der Füllstoffpartikel durch Poren und Risse innerhalb der monolithischen Proben teilweise kompensiert wird, wodurch sich die gesamte Ausdehnung im Vergleich zu dichten Proben verringert.

Da die durch die Oxidation des aktiven Füllstoffes $ZrSi_2$ verursachte Ausdehnung irreversibel ist und der TAK des Stahlsubstrats den der Beschichtung überschreitet, schrumpft der Stahl stärker als die Beschichtung während der Abkühlphase nach der Pyrolyse, wodurch sich die Druckspannungen in den Beschichtungen erhöhen. Diese Spannungen verhindern die Bildung von Segmentationsrissen – Risse quer durch die Schichtdicke – verursachen aber die Verbreiterung von diagonalen Rissen. Allerdings weist die Mikrostruktur hauptsächlich parallel zur Substratoberfläche orientierte Risse auf. Aufgrund der Expansion der aktiven Füllstoffpartikel, dehnt sich das Beschichtungsmaterial in alle Richtungen aus. Jedoch ist die Ausdehnung der Schicht parallel zur Substratoberfläche durch die Haftung mit dem Substrat eingeschränkt und kann nur in der Schichtdicke frei stattfinden. Aufgrund dieser Ausdehnung der Schichtdicke entstehen Zugspannungen, die zur Rissbildung parallel zur Substratoberfläche führen. Eine weitere Auslagerung der Beschichtungen bei hohen Temperaturen führt zu einer weiteren Oxidation des Füllstoffs, wodurch die Druckspannungen steigen. Nach ca. 10 h bei 1000 °C ist die Oxidation abgeschlossen und die Beschichtung dehnt sich nicht weiter aus.

Untersuchungen der Haftfestigkeit haben die Entwicklung von Beschichtungen mit starken Adhäsions- und Kohäsionskräften durch die Bildung kovalenter Bindungen zwischen dem Precursor und dem Substrat bzw. den Füllstoffpartikeln bestätigt. Um die Haftfestigkeit zu charakterisieren, wurden Gitterschnitttests durchgeführt, wobei sich die PDC-basierten Wärmedämmschichten durch eine visuelle Überprüfung in die Klasse Gt-1 (Abplatzung von maximal 5% der getesteten Fläche) einstufen lassen. Durch die Untersuchung mittels Lichtmikroskopie zeigte sich, dass der Kohäsionsbruch der dominierende Versagensmechanismus ist, der auf ca. 51% der getesteten Fläche stattfindet. Die Haftfestigkeit wurde außerdem mittels Abzugstests (Pull-Off) quantifiziert, woraus sich eine durchschnittliche Haftfestigkeit von 20,9 MPa ergab, die im Bereich der Haftfestigkeitswerte von konventionellen Wärmedämmschichten liegt. Bei den Tests wurden sowohl Adhäsions- als auch Kohäsionsbrüche beobachtet.

Die Messung der Wärmeleitfähigkeit erfolgte mittels $3\omega$-Methode. Aufgrund der geringen Wärmeleitfähigkeit der einzelnen Komponenten und der optimierten Mikrostruktur beträgt die Wärmeleitfähigkeit der entwickelten PDC-basierten Wärmedämmschichten nur $0,54 \pm 0,14$ W m$^{-1}$ K$^{-1}$ bei Raumtemperatur und $0,85 \pm 0,08$ W m$^{-1}$ K$^{-1}$ bei 500 °C. Diese Werte sind geringer als die durchschnittliche Wärmeleitfähigkeit von konventionellen Wärmedämmschichten (0.8-1,5 W m$^{-1}$ K$^{-1}$).

Die Beständigkeit der Wärmedämmschichten bei Temperaturwechseln und Oxidation wurde anschließend untersucht. Die Beschichtungen überstanden 30 Zyklen Temperaturwechsel mit einem Temperaturunterschied von ca. 1100 °C ohne Abplatzungen. Außerdem konnte nachgewiesen werden, dass das PHPS Bond-Coat äußerst wichtig für die Erhöhung der Oxidationsbeständigkeit des gesamten Systems ist. Nach einer Auslagerung bei 900 °C für 100 h an Luft wiesen die auf die Substrate ohne PHPS Schicht applizierten Wärmedämmschichten kaum Haftung auf und die Dicke der Oxidschicht auf dem Stahlsubstrat betrug ~5 µm. Im Gegensatz dazu blieben die auf die mit PHPS beschichten Substrate aufgetragen Wärmedämmschichten weiterhin stabil und die Dicke der thermisch gewachsenen Oxidschicht betrug lediglich ~2,5 µm. Während der Auslagerung bei 900 °C diffundieren Cr und Mn – und später auch Ti – durch die PHPS Schicht und bilden eine TGO zwischen dem PHPS Bond-Coat und der Wärmedämmschicht. Die PHPS Schicht wirkt dann wie ein Haftvermittler, der die Haftung der TGO am Substrat verstärkt. Allerdings findet diese Diffusion der metallischen Elemente ungleichmäßig statt, wodurch sich die Grenzfläche zwischen dem Substrat und dem Bond-Coat nach einer Auslagerung für 100 h deutlich änderte.

Außerdem wurde ein Verfahren für das Sprühbeschichten der Innenseite von Auspuffrohren im Rahmen der Doktorarbeit entwickelt. Rohre mit einer Länge von 500 mm und einem Innendurchmesser von 53 mm ließen sich dadurch mittels Sprühen beschichten. Trotz der geringen Schichtdicke im Vergleich zu konventionellen Wärmedämmschichten zeigten Untersuchungen hinsichtlich der thermischen Isolation, dass die PDC-basierten Wärmedämmschichten die Wärmeübertragung eines innerhalb des Rohrs fließenden Luftstromes zur Rohrwand verringerten. Aufgrund der geringen Wärmeleitfähigkeit der Beschichtung ließ sich die Wandtemperatur auf der Außenseite des Rohrs um 5 K im Vergleich zu nicht beschichteten Rohren reduzieren.

# 8 REFERENCES

[AbGu2004]  Abe, Y.; Gunji, T., **Oligo- and polysiloxanes,** *Prog. Polym. Sci.,* 29 [3], 2004; pp. 149–182.

[Acci2013]  Acciai Speciali Terni S.p.A. con Unico Socio, **Materials datasheet EN 1.4509: Stainless steel flat products,** 2013.

[Acro2012]  Acros Organics BVBA, **Materials safety datasheet: Di-n-butyl ether,** 2012.

[AeMe2004]  Aegerter, M.A.; Mennig, M. (Eds.), **Sol-gel technologies for glass producers and users,** Springer US, USA, 2004.

[AEOR1997]  Arfsten, N. J.; Eberle, A.; Otto, J.; Reich, A., **Investigations on the angle-dependent dip coating technique (ADDC) for the production of optical filters,** *J. Sol-Gel Sci. Technol.,* 8 [1/2/3], 1997; pp. 1099–1104.

[AFMS2014]  Amouzou, D.; Fourdrinier, L.; Maseri, F.; Sporken, R., **Formation of Me–O–Si covalent bonds at the interface between polysilazane and stainless steel,** *Appl. Surf. Sci.,* 320, 2014; pp. 519–523.

[AgTs2000a]  Agrafiotis, C.; Tsetsekou, A., **The effect of powder characteristics on washcoat quality: Part I - alumina washcoats,** *J. Eur. Ceram. Soc.,* 20 [7], 2000; pp. 815–824.

[AgTs2000b]  Agrafiotis, C.; Tsetsekou, A., **The effect of powder characteristics on washcoat quality: Part II - zirconia, titania washcoats — multilayered structures,** *J. Eur. Ceram. Soc.,* 20 [7], 2000; pp. 825–834.

[AiHe1960]  Ainger, F. W.; Herbert, J. M., **The preparation of phosphorus-nitrogen compounds as non-porous solids** in: *Special Ceramics* (Ed. Popper, P.). Academic Press, New York, 1960; pp. 168–181.

[AlEK2008]  Altun, O.; Erhan Boke, Y.; Kalemtas, A., **Problems for determining the thermal conductivity of TBCs by laser-flash method,** *JAMME,* 30 [2], 2008; pp. 115–120.

[APGA2004]  Aegerter, M. A.; Puetz, J.; Gasparro, G.; Al-Dahoudi, N., **Versatile wet deposition techniques for functional oxide coatings,** *Opt. Mater.,* 26 [2], 2004; pp. 155–162.

[ASPS2010]  Athanasopoulos, G. I.; Svoukis, E.; Pervolaraki, M.; Saint-Martin, R.; Revcolevschi, A.; Giapintzakis, J., **Thermal conductivity of Ni, Co, and Fe-doped La5Ca9Cu24O41 thin films measured by the 3ω method,** *Thin Solid Films,* 518 [16], 2010; pp. 4684–4687.

[ASTM2009]  ASTM D4541, D01 Committee, **Test method for pull-off strength of coatings using portable adhesion testers,** ASTM International, West Conshohocken, PA, 2009.

[ASTM2013]  ASTM C633, B08 Committee, **Test method for adhesion or cohesion strength of thermal spray coatings,** ASTM International, West Conshohocken, PA, 2013.

[AZEM2013]  AZ Electronic Materials (Merck KGaA), **Materials safety datasheet: KiON HTT 1800,** 2013.

[BaKM2015]   Barroso, G. S.; Krenkel, W.; Motz, G., **Low thermal conductivity coating system for application up to 1000°C by simple PDC processing with active and passive fillers,** *J. Eur. Ceram. Soc.,* 35 [12], 2015; pp. 3339–3348.

[BaSo1991]   Babonneau, F.; Sorarù, G. D., **Synthesis and characterization of Si-Zr-C-O ceramics from polymer precursors,** *J. Eur. Ceram. Soc.,* 8 [1], 1991; pp. 29–34.

[BDDE2005]   Bauer, F.; Decker, U.; Dierdorf, A.; Ernst, H.; Heller, R.; Liebe, H.; Mehnert, R., **Preparation of moisture curable polysilazane coatings. Part I: Elucidation of low temperature curing kinetics by FT-IR spectroscopy,** *Prog. Org. Coat.,* 53 [3], 2005; pp. 183–190.

[BEBC2016]   Biasetto, L.; Elsayed, H.; Bonollo, F.; Colombo, P., **Polymer-derived sphene biocoating on cpTi substrates for orthopedic and dental implants,** *Surf. Coat. Technol.,* 301, 2016; pp. 140–147.

[BeMS2004]   Berni, A.; Mennig, M.; Schmidt, H., **Doctor Blade** in: *Sol-gel technologies for glass producers and users* (Eds. Aegerter, M. A.; Mennig, M.). Springer US, USA, 2004; pp. 89–92.

[BeMv1999]   Beele, W.; Marijnissen, G.; van Lieshout, A., **The evolution of thermal barrier coatings: status and upcoming solutions for today's key issues,** *Surf. Coat. Technol.,* 120-121, 1999; pp. 61–67.

[Berg2015]   Berger, L.-M., **Application of hardmetals as thermal spray coatings,** *Int. J. Refract. Met. Hard Mater.,* 49, 2015; pp. 350–364.

[Bern2012]   Bernard, S. (Ed.), **Design, processing and properties of ceramic materials from preceramic precursors,** Nova Science Publishers Inc, New York, 2012.

[BFPS2014]   Bernardo, E.; Fiocco, L.; Parcianello, G.; Storti, E.; Colombo, P., **Advanced ceramics from preceramic polymers modified at the nano-scale: A review,** *Materials,* 7 [3], 2014; pp. 1927–1956.

[BGHS2007]   Bakumov, V.; Gueinzius, K.; Hermann, C.; Schwarz, M.; Kroke, E., **Polysilazane-derived antibacterial silver–ceramic nanocomposites,** *J. Eur. Ceram. Soc.,* 27 [10], 2007; pp. 3287–3292.

[BiAl1995]   Bill, J.; Aldinger, F., **Precursor-derived covalent ceramics,** *Adv. Mater.,* 7 [9], 1995; pp. 775–787.

[BiHe1996]   Bill, J.; Heimann, D., **Polymer-derived ceramic coatings on C/C-SiC composites,** *J. Eur. Ceram. Soc.,* 16 [10], 1996; pp. 1115–1120.

[Birn2004]   Birnie III, D. P., **Spin coating technique** in: *Sol-gel technologies for glass producers and users* (Eds. Aegerter, M. A.; Mennig, M.). Springer US, USA, 2004; pp. 49–55.

[BKDS2016]   Barroso, G.; Kraus, T.; Degenhardt, U.; Scheffler, M.; Motz, G., **Functional coatings based on preceramic polymers,** *Adv. Eng. Mater.,* 18 [5], 2016; pp. 746–753.

[BKMW2001]  Bill, J.; Kamphowe, T. W.; Müller, A.; Wichmann, T.; Zern, A.; Jalowieki, A.; Mayer, J.; Weinmann, M.; Schuhmacher, J.; Müller, K.; Peng, J.; Seifert, H. J.; Aldinger, F., **Precursor-derived Si-(B-)C-N ceramics: Thermolysis,** amorphous state and crystallization, *Appl. Organometal. Chem.*, 15 [10], 2001; pp. 777–793.

[BlMK2005]  Blum, Y. D.; MacQueen, D. B.; Kleebe, H.-J., **Synthesis and characterization of carbon-enriched silicon oxycarbides,** *J. Eur. Ceram. Soc.*, 25 [2-3], 2005; pp. 143–149.

[BlSL1989]  Blum, Y. D.; Schwartz, K. B.; Laine, R. M., **Preceramic polymer pyrolysis: Part 1, pyrolytic properties of polysilazanes,** *J. Mater. Sci.*, 24 [5], 1989; pp. 1707–1718.

[BMSR2007]  Bai, J.; Maute, K.; Shah, S. R.; Raj, R., **Mechanical Design for Accommodating Thermal Expansion Mismatch in Multilayer Coatings for Environmental Protection at Ultrahigh Temperatures,** *J. Am. Ceram. Soc.*, 90 [1], 2007; pp. 170–176.

[BoJa1993]  Bordia, R. K.; Jagota, A., **Crack growth and damage in constrained sintering films,** *J. Am. Ceram. Soc.*, 76 [10], 1993; pp. 2475–2485.

[BoRa1985]  Bordia, R. K.; Raj, R., **Sintering Behavior of Ceramic Films Constrained by a Rigid Substrate,** *J. Am. Ceram. Soc.*, 68 [6], 1985; pp. 287–292.

[BPGC1993]  Bahloul, D.; Pereira, M.; Goursat, P.; Choong Kwet Yive, N. S.; Corriu, R. J. P., **Preparation of silicon carbonitrides from an organosilicon polymer: I, Thermal decomposition of the cross-linked polysilazane,** *J. Am. Ceram. Soc.*, 76 [5], 1993; pp. 1156–1162.

[BrBo1963]  Brace, W. F.; Bombolakis, E. G., **A note on brittle crack growth in compression,** *J. Geophys. Res.*, 68 [12], 1963; pp. 3709–3713.

[Brin2013]  Brinker, C. J., **Dip coating** in: *Chemical solution deposition of functional oxide thin films* (Eds. Schneller, T. et al.). Springer Vienna, Vienna, 2013; pp. 233–261.

[BrLB1992]  Brotzen, F. R.; Loos, P. J.; Brady, D. P., **Thermal conductivity of thin SiO2 films,** *Thin Solid Films*, 207 [1-2], 1992; pp. 197–201.

[BrTa1991]  Brandon, J. R.; Taylor, R., **Phase stability of zirconia-based thermal barrier coatings part I. Zirconia-yttria alloys,** *Surf. Coat. Technol.*, 46 [1], 1991; pp. 75–90.

[BSMH2001]  Bouyer, E.; Schiller, G.; Müller, M.; Henne, R. H., **Thermal plasma chemical vapor deposition of Si-based ceramic coatings from liquid precursors,** *Plasma Chem. Plasma Process.*, 21 [4], 2001; pp. 523–546.

[BWAM2004]  Berger, F.; Weinmann, M.; Aldinger, F.; Müller, K., **Solid-state NMR studies of the preparation of Si–Al–C–N ceramics from aluminum-modified polysilazanes and polysilylcarbodiimides,** *Chem. Mater.*, 16 [5], 2004; pp. 919–929.

[Cahi1990]  Cahill, D. G., **Thermal conductivity measurement from 30 to 750 K: The 3ω method,** *Rev. Sci. Instrum.*, 61 [2], 1990; p. 802.

[CaJa2008]  Carslaw, H. S.; Jaeger, J. C., **Conduction of heat in solids,** 2. ed., reprinted. Edition, Clarendon Press, Oxford, 2008.

[CaPo1987]  Cahill, D. G.; Pohl, R. O., **Thermal conductivity of amorphous solids above the plateau,** *Phys. Rev. B*, 35 [8], 1987; pp. 4067–4073.

[CaVS2004]    Cao, X. Q.; Vassen, R.; Stoever, D., **Ceramic materials for thermal barrier coatings,** *J. Eur. Ceram. Soc.,* 24 [1], 2004; pp. 1–10.

[CBBM2015]    Chiavari, C.; Balbo, A.; Bernardi, E.; Martini, C.; Zanotto, F.; Vassura, I.; Bignozzi, M. C.; Monticelli, C., **Organosilane coatings applied on bronze: Influence of UV radiation and thermal cycles on the protectiveness,** *Prog. Org. Coat.,* 82, 2015; pp. 91–100.

[CBMS2015]    Coan, T.; Barroso, G. S.; Machado, R.A.F.; Souza, F. S. de; Spinelli, A.; Motz, G., **A novel organic-inorganic PMMA/polysilazane hybrid polymer for corrosion protection,** *Prog. Org. Coat.,* 89, 2015; pp. 220–230.

[CGVC2009]    Chevalier, J.; Gremillard, L.; Virkar, A. V.; Clarke, D. R., **The tetragonal-monoclinic transformation in zirconia: Lessons learned and future trends,** *J. Am. Ceram. Soc.,* 92 [9], 2009; pp. 1901–1920.

[ChKY1987]    Chou, Y. T.; Ko, Y. T.; Yan, M. F., **Fluid flow model for ceramic tape casting,** *J. Am. Ceram. Soc.,* 70 [10], 1987; C-280-C-282.

[ChMa1991]    Chandra, G.; Martin, T. E., **Rapid thermal process for obtaining silica coatings.** USA, Patent No. US 5059448 A, 1991.

[ChPo1964]    Chantrell, P. G.; Popper, P., **Inorganic polymers and ceramics** in: *Special Ceramics* (Ed. Popper, P.). Academic Press, New York, 1964; pp. 87–103.

[ChRa1989]    Cheng, T.; Raj, R., **Flaw generation during constrained sintering of metal-ceramic and metal-glass multilayer films,** *J. Am. Ceram. Soc.,* 72 [9], 1989; pp. 1649–1655.

[CiGC2009]    Cipitria, A.; Golosnoy, I. O.; Clyne, T. W., **A sintering model for plasma-sprayed zirconia TBCs. Part I: Free-standing coatings,** *Acta Mater.,* 57 [4], 2009; pp. 980–992.

[CJBG1998]    Cheng, J.; Jordan, E.H.; Barber, B.; Gell, M., **Thermal/residual stress in an electron beam physical vapor deposited thermal barrier coating system,** *Acta Mater.,* 46 [16], 1998; pp. 5839–5850.

[ClPh2005]    Clarke, D. R.; Phillpot, S. R., **Thermal barrier coating materials,** *Mater. Today,* 8 [6], 2005; pp. 22–29.

[CMFV2001]    Colombo, P.; Martucci, A.; Fogato, O.; Villoresi, P., **Silicon carbide films by laser pyrolysis of polycarbosilane,** *J. Am. Ceram. Soc.,* 84 [1], 2001; pp. 224–226.

[CMRS2010]    Colombo, P.; Mera, G.; Riedel, R.; Sorarù, G. D., **Polymer-derived ceramics: 40 years of research and innovation in advanced ceramics,** *J. Eur. Ceram. Soc.,* 93 [7], 2010; pp. 1805–1837.

[Coel2012]    Coeling, K. J., **Coating methods: Spray** in: *Processing and finishing of polymeric materials* (Ed. John Wiley & Sons, Inc). John Wiley & Sohns, USA, 2012; pp. 343–351.

[CoHe2002]    Colombo, P.; Hellmann, J., **Ceramic foams from preceramic polymers,** *Mater. Res. Innovations,* 6 [5-6], 2002; pp. 260–272.

[CoLo2008]    Corral, E. L.; Loehman, R. E., **Ultra-high-temperature ceramic coatings for oxidation protection of carbon–carbon composites,** *J. Am. Ceram. Soc.,* 91 [5], 2008; pp. 1495–1502.

[CRSK2010]  Colombo, P.; Riedel, R.; Sorarù, G.D.; Kleebe, H.-J. (Eds.), **Polymer derived ceramics: From nano-structure to applications,** DEStech Publ, Lancaster Pa., 2010.

[Cver2002]  Cverna, F., **ASM ready reference: thermal properties of metals,** 1st Edition, ASM International, Materials Park, Ohio, 2002.

[CWAT2007]  Chandra-Ambhorn, S.; Wouters, Y.; Antoni, L.; Toscan, F.; Galerie, A., **Adhesion of oxide scales grown on ferritic stainless steels in solid oxide fuel cells temperature and atmosphere conditions,** *J. Power Sources,* 171 [2], 2007; pp. 688–695.

[Deut2015]  Deutsche Edelstahlwerke GmbH, **Werkstoffdatenblatt: Nichtrostender ferritischer Chrom-Stahl 1.4509,** 2015.

[DIN1999]  DIN EN 10095:1999-05, **Heat resisting steels and nickel alloys; German version EN 10095:1999,** Beuth Verlag GmbH, 05/1999.

[DIN2013]  DIN EN ISO 2409:2013-06, **Paints and varnishes - Cross-cut test (ISO 2409:2013); German version EN ISO 2409:2013,** Beuth Verlag GmbH, 2013.

[DIN2014]  DIN EN 10088-1:2014-12, **Stainless steels - Part 1: List of stainless steels; German version EN 10088-1:2014,** Beuth Verlag GmbH, 12/2014.

[DrRi1997]  Dressler, W.; Riedel, R., **Progress in silicon-based non-oxide structural ceramics,** *Int. J. Refract. Met. Hard Mater.,* 15 [1-3], 1997; pp. 13–47.

[DSCP2000]  Danko, G. A.; Silberglitt, R.; Colombo, P.; Pippel, E.; Woltersdorf, J., **Comparison of microwave hybrid and conventional heating of preceramic polymers to form silicon carbide and silicon oxycarbide ceramics,** *J. Am. Ceram. Soc.,* 83 [7], 2000; pp. 1617–1625.

[EBJÖ2011]  Eriksson, R.; Brodin, H.; Johansson, S.; Östergren, L.; Li, X.-H., **Influence of isothermal and cyclic heat treatments on the adhesion of plasma sprayed thermal barrier coatings,** *Surf. Coat. Technol.,* 205 [23-24], 2011; pp. 5422–5429.

[EEA 2015]  EEA - European Environment Agency, **European Union emission inventory report 1990–2013 under the UNECE Convention on Long-range Transboundary Air Pollution (LRTAP),** Publications Office of the European Union, Luxembourg, 2015.

[ElST2007]  Elyassi, B.; Sahimi, M.; Tsotsis, T. T., **Silicon carbide membranes for gas separation applications,** *J. Membr. Sci.,* 288 [1-2], 2007; pp. 290–297.

[ElST2008]  Elyassi, B.; Sahimi, M.; Tsotsis, T. T., **A novel sacrificial interlayer-based method for the preparation of silicon carbide membranes,** *J. Membr. Sci.,* 316 [1-2], 2008; pp. 73–79.

[EmBP1958]  Emslie, A. G.; Bonner, F. T.; Peck, L. G., **Flow of a viscous liquid on a rotating disk,** *J. Appl. Phys.,* 29 [5], 1958; p. 858.

[Erja2000]  Erjavec, J., **Automotive technology: A systems approach,** 3rd Edition, Delmar, Albany NY u.a., 2000.

[Euro1970]  **Directive 70/220/EEC,** Official Journal of the European Communities, L76/1, 1970.

[Euro1991]    Directive 91/441/EEC, Official Journal of the European Communities, L242, 1991.

[Euro1996]    Directive 96/69/EC, Official Journal of the European Communities, L282/67, 1996.

[Euro1998]    Directive 98/69/EC, Official Journal of the European Communities, L350, 1998.

[Euro2002]    Directive 2001/116/EC, Official Journal of the European Communities, L18, 2002.

[Euro2008]    Regulation (EC) 692/2008, Official Journal of the European Union, L199, 2008.

[EvHH2001]    Evans, A. G.; He, M. Y.; Hutchinson, J. W., **Mechanics-based scaling laws for the durability of thermal barrier coatings**, Prog. Mater. Sci., 46 [3-4], 2001; pp. 249–271.

[EvHu1984]    Evans, A. G.; Hutchinson, J. W., **On the mechanics of delamination and spalling in compressed films**, Int. J. Solids Struct., 20 [5], 1984; pp. 455–466.

[FBNK2014]    Flores, O.; Bordia, R. K.; Nestler, D.; Krenkel, W.; Motz, G., **Ceramic fibers based on SiC and SiCN systems: Current research, development, and commercial status**, Adv. Eng. Mater., 16 [6], 2014; pp. 621–636.

[FrGS2001]    Friedrich, C.; Gadow, R.; Schirmer, T., **Lanthanum hexaaluminate — a new material for atmospheric plasma spraying of advanced thermal barrier coatings**, J. Therm. Spray Technol., 10 [4], 2001; pp. 592–598.

[FrRa1956]    Fritz, G.; Raabe, B., **Bildung siliciumorganischer Verbindungen: V. Die Thermische Zersetzung von Si(CH3)4 und Si(C2H5)4**, Z. Anorg. Allg. Chem., 286 [3-4], 1956; pp. 149–167.

[FSKH2013]    Flores, O.; Schmalz, T.; Krenkel, W.; Heymann, L.; Motz, G., **Selective cross-linking of oligosilazanes to tailored meltable polysilazanes for the processing of ceramic SiCN fibres**, J. Mater. Chem. A, 1 [48], 2013; p. 15406.

[FTNS2005]    Friedel, T.; Travitzky, N.; Niebling, F.; Scheffler, M.; Greil, P., **Fabrication of polymer derived ceramic parts by selective laser curing**, J. Eur. Ceram. Soc., 25 [2-3], 2005; pp. 193–197.

[GaNi2006]    Gadow, R.; Niessen, K. v., **Lightweight ballistic with additional stab protection made of thermally sprayed ceramic and cermet coatings on aramide fabrics**, Int. J. Appl. Ceram. Technol., 3 [4], 2006; pp. 284–292.

[GCTW2008]    Gonzales, S.; Combarmond, L.; Tran, M. T.; Wouters, Y.; Galerie, A., **Short term oxidation of stainless steels during final annealing**, Mater. Sci. Forum, 595-598, 2008; pp. 601–610.

[GFHT2004]    Goerke, O.; Feike, E.; Heine, T.; Trampert, A.; Schubert, H., **Ceramic coatings processed by spraying of siloxane precursors (polymer-spraying)**, J. Eur. Ceram. Soc., 24 [7], 2004; pp. 2141–2147.

[GJVM1999]    Gell, M.; Jordan, E.; Vaidyanathan, K.; McCarron, K.; Barber, B.; Sohn, Y.-H.; Tolpygo, V. K., **Bond strength, bond stress and spallation** mechanisms of thermal barrier coatings, *Surf. Coat. Technol.*, 120-121, 1999; pp. 53–60.

[GKDD2009]   Günthner, M.; Kraus, T.; Dierdorf, A.; Decker, D.; Krenkel, W.; Motz, G., **Advanced coatings on the basis of Si(C)N precursors for protection of steel against oxidation,** *J. Eur. Ceram. Soc.*, 29 [10], 2009; pp. 2061–2068.

[GKKM2009]   Günthner, M.; Kraus, T.; Krenkel, W.; Motz, G.; Dierdorf, A.; Decker, D., **Particle-filled PHPS silazane-based coatings on steel,** *Int. J. Appl. Ceram. Technol.*, 6 [3], 2009; pp. 373–380.

[GoCl1998]   Gong, X.-Y.; Clarke, D. R., **On the measurement of strain in coatings formed on a wrinkled elastic substrate,** *Oxid. Met.*, 50 [5/6], 1998; pp. 355–376.

[GPBH2008]   Geßwein, H.; Pfrengle, A.; Binder, J. R.; Haußelt, J., **Kinetic model of the oxidation of ZrSi2 powders,** *J. Therm. Anal. Calorim.*, 91 [2], 2008; pp. 517–523.

[GPKM2014]   Günthner, M.; Pscherer, M.; Kaufmann, C.; Motz, G., **High emissivity coatings based on polysilazanes for flexible Cu(In,Ga)Se2 thin-film solar cells,** *Sol. Energy Mater. Sol. Cells,* 123, 2014; pp. 97–103.

[Grei1995]   Greil, P., **Active-filler-controlled pyrolysis of preceramic polymers,** *J. Am. Ceram. Soc.*, 78 [4], 1995; pp. 835–848.

[Grei1998]   Greil, P., **Near net shape manufacturing of polymer derived ceramics,** *J. Eur. Ceram. Soc.*, 18 [13], 1998; pp. 1905–1914.

[Grei2000]   Greil, P., **Polymer derived engineering ceramics,** *Adv. Eng. Mater.*, 2 [6], 2000; pp. 339–348.

[GrFS2014]   Grolig, J. G.; Froitzheim, J.; Svensson, J.-E., **Coated stainless steel 441 as interconnect material for solid oxide fuel cells: Oxidation performance and chromium evaporation,** *J. Power Sources,* 248, 2014; pp. 1007–1013.

[GRGH2009]   Girotto, C.; Rand, B. P.; Genoe, J.; Heremans, P., **Exploring spray coating as a deposition technique for the fabrication of solution-processed solar cells,** *Sol. Energy Mater. Sol. Cells,* 93 [4], 2009; pp. 454–458.

[GrSe1991]   Greil, P.; Seibold, M., **Active-filler-controlled pyrolysis (AFCOP): A novel fabrication route to ceramic composite materials** in: *Advanced composite materials. Processing, microstructures, bulk, and interfacial properties, characterization methods, and applications* (Ed. Sacks, M. D.). American Ceramic Society, Westerville, Ohio, 1991.

[GrSe1992]   Greil, P.; Seibold, M., **Modelling of dimensional changes during polymer-ceramic conversion for bulk component fabrication,** *J. Mater. Sci.*, 27 [4], 1992; pp. 1053–1060.

[GRSM1997]   Gabriel, A. O.; Riedel, R.; Storck, S.; Maier, W. F., **Synthesis and thermally induced ceramization of a non-oxidic poly(methylsilsesquicarbodi-imide) gel,** *Appl. Organometal. Chem.*, 11 [10-11], 1997; pp. 833–841.

[GRST1997]   Gaskell, P. H.; Rand, B.; Summers, J. L.; Thompson, H. M., **The effect of reservoir geometry on the flow within ceramic tape casters,** *J. Eur. Ceram. Soc.*, 17 [10], 1997; pp. 1185–1192.

[GSGW2011]  Günthner, M.; Schütz, A.; Glatzel, U.; Wang, K.; Bordia, R. K.; Greißl, O.; Krenkel, W.; Motz, G., **High performance environmental barrier coatings: Part I, Passive filler loaded SiCN system for steel,** *J. Eur. Ceram. Soc.,* 31 [15], 2011; pp. 3003–3010.

[GSKH2010]  Glatz, G.; Schmalz, T.; Kraus, T.; Haarmann, F.; Motz, G.; Kempe, R., **Copper-containing SiCN precursor ceramics (Cu@SiCN) as selective hydrocarbon oxidation catalysts using air as an oxidant,** *Chem.-Eur. J.,* 16 [14], 2010; pp. 4231–4238.

[GTDM2004]  Galerie, A.; Toscan, F.; Dupeux, M.; Mougin, J.; Lucazeau, G.; Valot, C.; Huntz, A.-M.; Antoni, L., **Stress and adhesion of chromia-rich scales on ferritic stainless steels in relation with spallation,** *Mat. Res.,* 7 [1], 2004; pp. 81–88.

[Hase1992]  Hasegawa, Y., **Si-C fiber prepared from polycarbosilane cured without oxygen,** *J. Inorg. Organomet. Polym.,* 2 [1], 1992; pp. 161–169.

[Haun2013]  Hauner Metallische Werkstoffe, **Materials safety datasheet: Zirconium disilicide,** 2013.

[HBFL1999]  Hsueh, C. H.; Becher, P. F.; Fuller, E. R.; Langer, S. A.; Carter, W. C., **Surface-roughness induced residual stresses in thermal barrier coatings: Computer simulations,** *Mater. Sci. Forum,* 308-311, 1999; pp. 442–449.

[HeCa1989]  Hemminger, W. F.; Cammenga, H. K., **Methoden der thermischen Analyse,** Springer, Berlin, 1989.

[HeEH2000]  He, M. Y.; Evans, A. G.; Hutchinson, J. W., **The ratcheting of compressed thermally grown thin films on ductile substrates,** *Acta Mater.,* 48 [10], 2000; pp. 2593–2601.

[HeSa1996]  Herman, H.; Sampath, S., **Thermal spray coatings** in: *Metallurgical and ceramic protective coatings* (Ed. Stern, K. H.). Chapman & Hall, London, 1996; pp. 261–289.

[HHRW1999] Hennige, V. D.; Haußelt, J.; Ritzhaupt-Kleissl, H.-J.; Windmann, T., **Shrinkage-free ZrSiO4-ceramics: Characterisation and applications,** *J. Eur. Ceram. Soc.,* 19 [16], 1999; pp. 2901–2908.

[HiTh2012]  Hillier, V. A. W.; Thornes, N., **Hillier's fundamentals of motor vehicle technology,** 6th Edition, Nelson Thornes, Cheltenham, 2012.

[HoBB1947]  Howatt, G. N.; Breckenridge, R. G.; Brownlow, J. M., **Fabrication of thin ceramic sheets for capacitors,** *J. Am. Ceram. Soc.,* 30 [8], 1947; pp. 237–242.

[HSBG2007]  Henager, C. H.; Shin, Y.; Blum, Y.; Giannuzzi, L. A.; Kempshall, B. W.; Schwarz, S. M., **Coatings and joining for SiC and SiC-composites for nuclear energy systems,** *J. Nucl. Mater.,* 367-370, 2007; pp. 1139–1143.

[HsFu2000]  Hsueh, C.H.; Fuller, E.R., **Analytical modeling of oxide thickness effects on residual stresses in thermal barrier coatings,** *Scripta Mater.,* 42 [8], 2000; pp. 781–787.

[HuXS2009]  Hu, C.; Xu, G.; Shen, X., **Preparation and characteristics of thermal resistance polysiloxane/Al composite coatings with low infrared emissivity,** *J. Alloys Compd.,* 486 [1-2], 2009; pp. 371–375.

[IcTI1987]     Ichikawa, H.; Teranishi, H.; Ishikawa, T., **Effect of curing conditions on mechanical properties of SiC fibre (Nicalon)**, *J. Mater. Sci. Lett.*, 6 [4], 1987; pp. 420–422.

[IGBA2002]     Ishihara, S.; Gu, H.; Bill, J.; Aldinger, F.; Wakai, F., **Densification of precursor-derived Si-C-N ceramics by high-pressure hot isostatic pressing**, *J. Am. Ceram. Soc.*, 85 [7], 2002; pp. 1706–1712.

[IMCP2012]     Issartel, J.; Martoia, S.; Charlot, F.; Parry, V.; Parry, G.; Estevez, R.; Wouters, Y., **High temperature behavior of the metal/oxide interface of ferritic stainless steels**, *Corros. Sci.*, 59, 2012; pp. 148–156.

[INOS2001]     Idesaki, A.; Narisawa, M.; Okamura, K.; Sugimoto, M.; Morita, Y.; Seguchi, T.; Itoh, M., **Application of electron beam curing for silicon carbide fiber synthesis from blend polymer of polycarbosilane and polyvinylsilane**, *Radiat. Phys. Chem.*, 60 [4-5], 2001; pp. 483–487.

[ISEA1999]     Inoue, H.; Sekizawa, K.; Eguchi, K.; Arai, H., **Thick-film coating of hexaaluminate catalyst on ceramic substrates for high-temperature combustion**, *Catal. Today*, 47 [1-4], 1999; pp. 181–190.

[ISSF2007]     ISSF - International Stainless Steel Forum, **Die ferritische Lösung: Eigenschaften, Vorteile, Einsatzmöglichkeiten - was Sie über ferritische nichtrostende Stähle wissen müssen**, Brussels, 2007.

[ISTN2004]     Idesaki, A.; Sugimoto, M.; Tanaka, S.; Narisawa, M.; Okamura, K.; Itoh, M., **Synthesis of a minute SiC product from polyvinylsilane with radiation curing Part I Radiation curing of polyvinylsilane**, *J. Mater. Sci.*, 39 [18], 2004; pp. 5689–5694.

[IZYI2004]     Ichiki, M.; Zhang, L.; Yang, Z.; Ikehara, T.; Maeda, R., **Thin film formation on non-planar surface with use of spray coating fabrication**, *Microsys. Technol.*, 10 [5], 2004; pp. 360–363.

[JaCS2010]     Jablonski, P. D.; Cowen, C. J.; Sears, J. S., **Exploration of alloy 441 chemistry for solid oxide fuel cell interconnect application**, *J. Power Sources*, 195 [3], 2010; pp. 813–820.

[JaHu1990]     Jagota, A.; Hui, C. Y., **Mechanics of sintering thin films — I: Formulation and analytical results**, *Mech. Mater.*, 9 [2], 1990; pp. 107–119.

[JaHu1991]     Jagota, A.; Hui, C. Y., **Mechanics of sintering thin films — II: Cracking due to self-stress**, *Mech. Mater.*, 11 [3], 1991; pp. 221–234.

[JSLF2012]     Jung, S.; Seo, D.; Lombardo, S. J.; Feng, Z. C.; Chen, J. K.; Zhang, Y., **Fabrication using filler controlled pyrolysis and characterization of polysilazane PDC RTD arrays on quartz wafers**, *Sens. Actuators, A*, 175, 2012; pp. 53–59.

[KeDe2008]     Kelly, J. R.; Denry, I., **Stabilized zirconia as a structural ceramic: An overview**, *Dent. Mater.*, 24 [3], 2008; pp. 289–298.

[KGKM2009]     Kraus, T.; Günthner, M.; Krenkel, W.; Motz, G., **cBN particle filled SiCN precursor coatings**, *Adv. Appl. Ceram.*, 108 [8], 2009; pp. 476–482.

[KiLl1901]     Kipping, F. S.; Lloyd, L. L., **Organic derivatives of silicon. Triphenylsilicol and alkyloxysilicon chlorides**, *J. Chem. Soc., Trans.*, 79, 1901; p. 449.

[Kinl1987]    Kinloch, A. J., **Adhesion and Adhesives: Science and Technology,** Springer Netherlands, Dordrecht, s.l., 1987.

[Kreb2009]    Krebs, F. C., **Fabrication and processing of polymer solar cells: A review of printing and coating techniques,** Sol. Energy Mater. Sol. Cells, 93 [4], 2009; pp. 394–412.

[KrUl2006]    Krüger, U.; Ullrich, R., **Producing a ceramic layer by spraying polymer ceramic precursor particles onto a surface comprises using a cold gas spray nozzle.** Germany, Patent No. DE 102005031101 B3, 2006.

[KuKa2016]    Kumar, V.; Kandasubramanian, B., **Processing and design methodologies for advanced and novel thermal barrier coatings for engineering applications,** Particuology, 27, 2016; pp. 1–28.

[KYWK1996]    Kan, C.; Yuan, Q.; Wang, M.; Kong, X., **Synthesis of silicone–acrylate copolymer latexes and their film properties,** Polym. Adv. Technol., 7 [2], 1996; pp. 95–97.

[LaBa1993]    Laine, R. M.; Babonneau, F., **Preceramic polymer routes to silicon carbide,** Chem. Mater., 5 [3], 1993; pp. 260–279.

[LaLe1942]    Landau, L.; Levich, B., **Dragging of a liquid by a moving plate,** Acta Physicochimica U.R.S.S., XVII [1-2], 1942; pp. 42–54.

[LaRe1997]    Larson, R. G.; Rehg, T. J., **Spin coating** in: Liquid film coating. Scientific principles and their technological implications (Ed. Kistler, S. F.). Chapman & Hall, London, 1997; pp. 709–734.

[LCGP2005]    Lin, M.; Chu, F.; Guyot, A.; Putaux, J.-L.; Bourgeat-Lami, E., **Silicone–polyacrylate composite latex particles: Particles formation and film properties,** Polymer, 46 [4], 2005; pp. 1331–1337.

[LiCl1996]    Lipkin, D. M.; Clarke, D. R., **Measurement of the stress in oxide scales formed by oxidation of alumina-forming alloys,** Oxid. Met., 45 [3-4], 1996; pp. 267–280.

[LiKM1996]    Li, Z.; Kusakabe, K.; Morooka, S., **Preparation of thermostable amorphous Si-C-O membrane and its application to gas separation at elevated temperature,** J. Membr. Sci., 118 [2], 1996; pp. 159–168.

[LiKM1997]    Li, Z.; Kusakabe, K.; Morooka, S., **Pore structure and permeance of amorphous Si-C-O membranes with high durability at elevated temperature,** Sep. Sci. Technol., 32 [7], 1997; pp. 1233–1254.

[Lipo1988]    Lipowitz, J., **Infusible preceramic polymers via plasma treatment.** USA, Patent No. US 4743662 A, 1988.

[LlCa2001]    Lloyd, A. C.; Cackette, T. A., **Diesel Engines: Environmental impact and control,** J. Air Waste Manage. Assoc., 51 [6], 2001; pp. 809–847.

[LNBC1993]    Labrousse, M.; Nanot, M.; Boch, P.; Chassagneux, E., **Ex-polymer SiC coatings with Al2O3 particulates as filler materials,** Ceram. Int., 19 [4], 1993; pp. 259–267.

[LPPV2003]    Lehmann, H.; Pitzer, D.; Pracht, G.; Vassen, R.; Stöver, D., **Thermal conductivity and thermal expansion coefficients of the lanthanum rare-**

earth-element zirconate system, *J. Am. Ceram. Soc.*, 86 [8], 2003; pp. 1338–1344.

[Luka2007]   Lukacs, A., **Polysilazane precursors to advanced ceramics**, *Am. Ceram. Soc. Bull.*, 86, 2007; pp. 9301–9306.

[LZHY2013]   Liu, J.; Zhang, L.; Hu, F.; Yang, J.; Cheng, L.; Wang, Y., **Polymer-derived yttrium silicate coatings on 2D C/SiC composites**, *J. Eur. Ceram. Soc.*, 33 [2], 2013; pp. 433–439.

[LZLC2010]   Liu, J.; Zhang, L.; Liu, Q.; Cheng, L.; Wang, Y., **Polymer-derived SiOC-barium-strontium aluminosilicate coatings as an environmental barrier for C/SiC composites**, *J. Am. Ceram. Soc.*, 93 [12], 2010; pp. 4148–4152.

[LZYC2012]   Liu, J.; Zhang, L.; Yang, J.; Cheng, L.; Wang, Y., **Fabrication of SiCN–Sc2Si2O7 coatings on C/SiC composites at low temperatures**, *J. Eur. Ceram. Soc.*, 32 [3], 2012; pp. 705–710.

[MBCT2007]   Miele, P.; Bernard, S.; Cornu, D.; Toury, B., **Recent developments in polymer-derived ceramic fibers (PDCFs): Preparation, properties and applications - A review**, *Soft Mater.*, 4 [2-4], 2007; pp. 249–286.

[MGDR2002]   Mougin, J.; Galerie, A.; Dupeux, M.; Rosman, N.; Lucazeau, G.; Huntz, A.-M.; Antoni, L., **In-situ determination of growth and thermal stresses in chromia scales formed on a ferritic stainless steel**, *Mater. Corros.*, 53 [7], 2002; pp. 486–490.

[MGXC2006]   Ma, W.; Gong, S.; Xu, H.; Cao, X., **The thermal cycling behavior of lanthanum–cerium oxide thermal barrier coating prepared by EB–PVD**, *Surf. Coat. Technol.*, 200 [16-17], 2006; pp. 5113–5118.

[MHLG2002]   Moon, S.; Hatano, M.; Lee, M.; Grigoropoulos, C. P., **Thermal conductivity of amorphous silicon thin films**, *Int. J. Heat Mass Transfer*, 45 [12], 2002; pp. 2439–2447.

[MHTA2003]   Müller, A.; Herlin-Boime, N.; Ténégal, F.; Armand, X.; Berger, F.; Flank, A. M.; Dez, R.; Müller, K.; Bill, J.; Aldinger, F., **Comparison of Si/C/N pre-ceramics obtained by laser pyrolysis or furnace thermolysis**, *J. Eur. Ceram. Soc.*, 23 [1], 2003; pp. 37–46.

[Mill1997]   Miller, R. A., **Thermal barrier coatings for aircraft engines: History and directions**, *J. Therm. Spray Technol.*, 6 [1], 1997; pp. 35–42.

[MJMP2008]   Ma, W.; Jarligo, M. O.; Mack, D. E.; Pitzer, D.; Malzbender, J.; Vaßen, R.; Stöver, D., **New generation perovskite thermal barrier coating materials**, *J. Therm. Spray Technol.*, 17 [5-6], 2008; pp. 831–837.

[MoSa2004]   Mori, Y.; Saito, R., **Synthesis of a poly(methyl methacrylate)/silica nano-composite by soaking of a microphase separated polymer film into a perhydropolysilazane solution**, *Polymer*, 45 [1], 2004; pp. 95–100.

[Motz2006]   Motz, G., **Synthesis of SiCN-precursors for fibres and matrices**, *Adv. Sci. Technol.*, 50, 2006; pp. 24–30.

[MRRQ2011]   Maleki, M.; Reyssat, M.; Restagno, F.; Quéré, D.; Clanet, C., **Landau-Levich menisci**, *J. Colloid Interface Sci.*, 354 [1], 2011; pp. 359–363.

[MSTK2012]   Motz, G.; Schmalz, T.; Trassl, S.; Kempe, R., **Oxidation behavior of SiCN materials** in: *Design, processing and properties of ceramic materials from* *preceramic precursors* (Ed. Bernard, S.). Nova Science Publishers Inc, New York, 2012.

[MuEv2000]   Mumm, D. R.; Evans, A. G., **On the role of imperfections in the failure of a thermal barrier coating made by electron beam deposition**, *Acta Mater.*, 48 [8], 2000; pp. 1815–1827.

[MUKS2006]   Mori, Y.; Ueda, T.; Kitaoka, S.; Sugahara, Y., **Preparation of Si-Al-N-C ceramic composites by yprolysis of blended precursors**, *J. Ceram. Soc. Japan*, 114 [1330], 2006; pp. 497–501.

[Niih1991]   Niihara, K., **New design concept of structural ceramics**, *J. Ceram. Soc. Japan*, 99 [1154], 1991; pp. 974–982.

[NoGL2005]   Norrman, K.; Ghanbari-Siahkali, A.; Larsen, N. B., **Studies of spin-coated polymer films**, *Annu. Rep. Prog. Chem., Sect. C*, 101, 2005; p. 174.

[OCCA1987]   OCCAA, Oil and Colours Chemists' Association Australia, **Surface coatings: Vol. 2 - Paints and their applications**, 2nd Edition, Tafe Educational Books, Australia, 1987.

[OKWB2011]   Ohl, C.; Kappa, M.; Wilker, V.; Bhattacharjee, S.; Scheffler, F.; Scheffler, M., **Novel open-cellular glass foams for optical applications**, *J. Am. Ceram. Soc.*, 94 [2], 2011; pp. 436–441.

[OlSB2001]   Olding, T.; Sayer, M.; Barrow, D., **Ceramic sol–gel composite coatings for electrical insulation**, *Thin Solid Films*, 398-399, 2001; pp. 581–586.

[Outo2012]   Outokumpu Group, **Outokumpu ferritic stainless steels**, 2012.

[Outo2016]   Outokumpu Group, **Materials datasheet: high temperature austenitic stainless steel**, 2016.

[PaGJ2002]   Padture, N. P.; Gell, M.; Jordan, E. H., **Thermal barrier coatings for gas-turbine engine applications**, *Science*, 296 [5566], 2002; pp. 280–284.

[PBCL2014]   Parcianello, G.; Bernardo, E.; Colombo, P.; Lenčéš, Z.; Vetrecín, M.; Šajgalík, P.; Kašiarová, M.; Soraru, G., **Preceramic polymer-derived SiAlON as sintering aid for silicon nitride**, *J. Am. Ceram. Soc.*, 97 [11], 2014; pp. 3407–3412.

[PeVB1990]   Peuckert, M.; Vaahs, T.; Brück, M., **Ceramics from organometallic polymers**, *Adv. Mater.*, 2 [9], 1990; pp. 398–404.

[PFBO2008]   Pavese, M.; Fino, P.; Badini, C.; Ortona, A.; Marino, G., **HfB2/SiC as a protective coating for 2D Cf/SiC composites: Effect of high temperature oxidation on mechanical properties**, *Surface and Coatings Technology*, 202 [10], 2008; pp. 2059–2067.

[PFSK1997]   Peters, M.; Fritscher, K.; Staniek, G.; Kaysser, W. A.; Schulz, U., **Design and properties of thermal barrier coatings for advanced turbine engines**, *Mat.-wiss. u. Werkstofftech.*, 28 [8], 1997; pp. 357–362.

[PGDM2008]   Perale, G.; Giordano, C.; Daniele, F.; Masi, M.; Colombo, P.; Gottardo, L.; Maccagnan, S., **A novel process for the manufacture of ceramic microelectrodes for biomedical applications**, *Int. J. Appl. Ceram. Technol.*, 5 [1], 2008; pp. 37–43.

[PiCo1997a]  Pivin, J. C.; Colombo, P., **Ceramic coatings by ion irradiation of polycarbosilanes and polysiloxanes, Part I: conversion mechanism,** *J. Mater. Sci.*, 32 [23], 1997; pp. 6163–6173.

[PiCo1997b]  Pivin, J. C.; Colombo, P., **Ceramic coatings by ion irradiation of polycarbosilanes and polysiloxanes, Part II: hardness and thermochemical stability,** *J. Mater. Sci.*, 32 [23], 1997; pp. 6175–6182.

[PiCS2000]  Pivin, J. C.; Colombo, P.; Sorarù, G. D., **Comparison of ion irradiation effects in silicon-based preceramic thin films,** *J. Am. Ceram. Soc.*, 83 [4], 2000; pp. 713–720.

[PiCT1996]  Pivin, J. C.; Colombo, P.; Tonidandel, M., **Ion irradiation of preceramic polymer thin films,** *J. Am. Ceram. Soc.*, 79 [7], 1996; pp. 1967–1970.

[PKKS2007]  Pham, T. A.; Kim, P.; Kwak, M.; Suh, K. Y.; Kim, D.-P., **Inorganic polymer photoresist for direct ceramic patterning by photolithography,** *Chem. Commun.* [39], 2007; pp. 4021–4023.

[PPFG2015]  Picard, L.; Phalip, P.; Fleury, E.; Ganachaud, F., **Chemical adhesion of silicone elastomers on primed metal surfaces: A comprehensive survey of open and patent literatures,** *Prog. Org. Coat.*, 80, 2015; pp. 120–141.

[PPWC2012]  Pan, W.; Phillpot, S. R.; Wan, C.; Chernatynskiy, A.; Qu, Z., **Low thermal conductivity oxides,** *MRS Bull.*, 37 [10], 2012; pp. 917–922.

[PuAe2004]  Puetz, J.; Aegerter, M. A., **Dip coating technique** in: *Sol-gel technologies for glass producers and users* (Eds. Aegerter, M. A.; Mennig, M.). Springer US, USA, 2004; pp. 37–48.

[PWMM2016]  Pradhan, D.; Wren, A. W.; Misture, S. T.; Mellott, N. P., **Investigating the structure and biocompatibility of niobium and titanium oxides as coatings for orthopedic metallic implants,** *Mater. Sci. Eng., C*, 58, 2016; pp. 918–926.

[RaEv2000]  Rabiei, A.; Evans, A. G., **Failure mechanisms associated with the thermally grown oxide in plasma-sprayed thermal barrier coatings,** *Acta Mater.*, 48 [15], 2000; pp. 3963–3976.

[RaLL1990]  Rabe, J. A.; Lipowitz, J.; Lu, P. P., **Curing preceramic polymers by exposure to nitrogen dioxide.** 1990, Patent No. US 5051215 A, 1991.

[Reif2014]  Reif, K. (Ed.), **Diesel Engine Management: Systems and Components,** Springer Fachmedien Wiesbaden, Wiesbaden, s.l., 2014.

[Reif2015]  Reif, K. (Ed.), **Gasoline Engine Management: Systems and Components,** Springer Fachmedien Wiesbaden, Wiesbaden, s.l., 2015.

[RGBB2005]  Rocha, R. M.; Greil, P.; Bressiani, J. C.; Bressiani, A. H. de A., **Complex-shaped ceramic composites obtained by machining compact polymer-filler mixtures,** *Mat. Res.*, 8 [2], 2005; pp. 191–196.

[RGMD1997]  Riedel, R.; Greiner, A.; Miehe, G.; Dressler, W.; Fuess, H.; Bill, J.; Aldinger, F., **The first crystalline solids in the ternary Si-C-N system,** *Angew. Chem. Int. Ed. Engl.*, 36 [6], 1997; pp. 603–606.

[RGWD2008]  Rufner, J.; Gannon, P.; White, P.; Deibert, M.; Teintze, S.; Smith, R.; Chen, H., **Oxidation behavior of stainless steel 430 and 441 at 800°C in single** (air/air) and dual atmosphere (air/hydrogen) exposures, *Int. J. Hydrogen Energy*, 33 [4], 2008; pp. 1392–1398.

[RKDR1996] Riedel, R.; Kienzle, A.; Dressler, W.; Ruwisch, L.; Bill, J.; Aldinger, F., **A silicoboron carbonitride ceramic stable to 2,000°C**, *Nature*, 382 [6594], 1996; pp. 796–798.

[RKGG1998] Riedel, R.; Kroke, E.; Greiner, A.; Gabriel, A. O.; Ruwisch, L.; Nicolich, J.; Kroll, P., **Inorganic solid-state chemistry with main group element carbodiimides**, *Chem. Mater.*, 10 [10], 1998; pp. 2964–2979.

[RKSA1995] Riedel, R.; Kleebe, H.-J.; Schönfelder, H.; Aldinger, F., **A covalent micro/nano-composite resistant to high-temperature oxidation**, *Nature*, 374 [6522], 1995; pp. 526–528.

[RMHK2006] Riedel, R.; Mera, G.; Hauser, R.; Klonczynski, A., **Silicon-based polymer-derived ceramics: Synthesis properties and applications - A review**, *J. Ceram. Soc. Japan*, 114 [1330], 2006; pp. 425–444.

[RRBB1997] Richter, R.; Roewer, G.; Böhme, U.; Busch, K.; Babonneau, F.; Martin, H. P.; Müller, E., **Organosilicon Polymers: Synthesis, architecture, reactivity and applications**, *Appl. Organometal. Chem.*, 11 [2], 1997; pp. 71–106.

[SBGH2004] Schulz, M.; Börner, M.; Göttert, J.; Hanemann, T.; Haußelt, J.; Motz, G., **Cross Linking Behavior of Preceramic Polymers Effected by UV- and Synchrotron Radiation**, *Adv. Eng. Mater.*, 6 [8], 2004; pp. 676–680.

[ScBG1994] Schwab, S. T.; Blanchard, C. R.; Graef, R. C., **The influence of preceramic binders on the microstructural development of silicon nitride**, *J. Mater. Sci.*, 29 [23], 1994; pp. 6320–6328.

[ScGa1985] Scherer, G. W.; Garino, T., **Viscous sintering on a rigid substrate**, *J. Am. Ceram. Soc.*, 68 [4], 1985; pp. 216–220.

[SCGP2009] Srisrual, A.; Coindeau, S.; Galerie, A.; Petit, J.-P.; Wouters, Y., **Identification by photoelectrochemistry of oxide phases grown during the initial stages of thermal oxidation of AISI 441 ferritic stainless steel in air or in water vapour**, *Corros. Sci.*, 51 [3], 2009; pp. 562–568.

[Schm1994] Schmidt, P. F., **Praxis der Rasterelektronenmikroskopie und Mikrobereichsanalyse**, expert-Verl., Renningen-Malmsheim, 1994.

[ScPK2001] Schlichting, K. W.; Padture, N. P.; Klemens, P. G., **Thermal conductivity of dense and porous yttria-stabilized zirconia**, *J. Mater. Sci.*, 36 [12], 2001; pp. 3003–3010.

[ScRE2012] Schneider, C. A.; Rasband, W. S.; Eliceiri, K. W., **NIH Image to ImageJ: 25 years of image analysis**, *Nat. Meth.*, 9 [7], 2012; pp. 671–675.

[ScRo1986] Schwartz, K. B.; Rowcliffe, D. J., **Modeling density contributions in preceramic polymer/ceramic powder systems**, *J. Am. Ceram. Soc.*, 69 [5], 1986; C-106-C-108.

[ScTo2010] Scheffler, F.; Torrey, J. D., **Coatings** in: *Polymer derived ceramics. From nanostructure to applications* (Eds. Colombo, P. et al.). DEStech Publ, Lancaster Pa., 2010; pp. 358–368.

[Seif2016] Seifert, M., **Entwicklung von Multiphasen-Verbundwerkstoffen im System Nb-Si-C-N auf Basis partikelgefüllter Polysilazane**, 1. Auflage Edition, Cuvillier Verlag, Göttingen, Niedersachs, 2016.

[SeWi1984] Seyferth, D.; Wiseman, G. H., **High-yield synthesis of Si3N4/SiC ceramic materials by pyrolysis of a novel polyorganosilazane**, *J. Am. Ceram. Soc.*, 67 [7], 1984; C-132-C-133.

[Seyf1995] Seyferth, D., **Preceramic polymers: Past, present and future** in: *Materials chemistry. An emerging discipline* (Eds. Interrante, L. V.; Casper, L. A.; Ellis, A. B.). American Chem. Soc, Washington DC, 1995; pp. 131–160.

[SGJM2016] Seifert, M.; Gonçalves, P.; Justus, T.; Martins, N.; Klein, A. N.; Motz, G., **A novel approach to develop composite ceramics based on active filler loaded precursor employing plasma assisted pyrolysis**, *Mater. Des.*, 89, 2016; pp. 893–900.

[SGMG2012] Schütz, A.; Günthner, M.; Motz, G.; Greißl, O.; Glatzel, U., **Characterisation of novel precursor-derived ceramic coatings with glass filler particles on steel substrates**, *Surf. Coat. Technol.*, 207, 2012; pp. 319–327.

[ShMc2000] Shelef, M.; McCabe, R.W., **Twenty-five years after introduction of automotive catalysts: What next?**, *Catal. Today*, 62 [1], 2000; pp. 35–50.

[SKJG2001] Sohn, Y.H.; Kim, J.H.; Jordan, E.H.; Gell, M., **Thermal cycling of EB-PVD/MCrAlY thermal barrier coatings: I. Microstructural development and spallation mechanisms**, *Surf. Coat. Technol.*, 146-147, 2001; pp. 70–78.

[SKKK2006] Song, I.-H.; Kim, M.-J.; Kim, H.-D.; Kim, Y.-W., **Processing of microcellular cordierite ceramics from a preceramic polymer**, *Scripta Mater.*, 54 [8], 2006; pp. 1521–1525.

[SoBM1988] Sorarù, G. D.; Babonneau, F.; Mackenzie, J. D., **Structural concepts on new amorphous covalent solids**, *J. Non-Cryst. Solids*, 106 [1-3], 1988; pp. 256–261.

[SoWa1999] Soliman, H. M.; Waheed, A. F., **Effect of differential thermal expansion coefficient on stresses generated in coating**, *J. Mater. Sci. Technol. (Journal of Materials Sciences and Technology)*, 15 [05], 1999; pp. 457–462.

[SPJG2003] Schlichting, K.W.; Padture, N.P.; Jordan, E.H.; Gell, M., **Failure modes in plasma-sprayed thermal barrier coatings**, *Mater. Sci. Eng., A*, 342 [1-2], 2003; pp. 120–130.

[SRCB1991] Sorarù, G. D.; Ravagni, A.; Campostrini, R.; Babonneau, F., **Synthesis and characterization of beta'-SiAlON ceramics from organosilicon polymers**, *J. Am. Ceram. Soc.*, 74 [9], 1991; pp. 2220–2223.

[SSBP2000] Schumann, E.; Sarioglu, C.; Blachere, J. R.; Pettit, F. S.; Meier, G. H., **High-temperature stress measurements during the oxidation of NiAl**, *Oxid. Met.*, 53 [3/4], 2000; pp. 259–272.

[SSKM1998] Suresh, G.; Seenivasan, G.; Krishnaiah, M.V.; Murti, P. S., **Investigation of the thermal conductivity of selected compounds of lanthanum, samarium and europium**, *J. Alloys Compd.*, 269 [1-2], 1998; pp. L9-L12.

[SSKM2015]   Seifert, M.; Shen, Z.; Krenkel, W.; Motz, G., **Nb(Si,C,N) composite materials densified by spark plasma sintering**, *J. Eur. Ceram. Soc.*, 35 [12], 2015; pp. 3319–3327.

[Ster1996]   Stern, K.H. (Ed.), **Metallurgical and ceramic protective coatings**, Chapman & Hall, London, 1996.

[StHS2009]   Stern, E.; Heyder, M.; Scheffler, F., **Micropatterned ceramic surfaces by coating with filled preceramic polymers**, *J. Am. Ceram. Soc.*, 92 [10], 2009; pp. 2438–2442.

[STKM2014]   Seifert, M.; Travitzky, N.; Krenkel, W.; Motz, G., **Multiphase ceramic composites derived by reaction of Nb and SiCN precursor**, *J. Eur. Ceram. Soc.*, 34 [8], 2014; pp. 1913–1921.

[Ston1909]   Stoney, G. G., **The tension of metallic films deposited by electrolysis**, *Proc. R. Soc. London, Ser. A*, 82 [553], 1909; pp. 172–175.

[Stra1985]   Strangman, T. E., **Thermal barrier coatings for turbine airfoils**, *Thin Solid Films*, 127 [1-2], 1985; pp. 93–106.

[SYUF2006a]  Suda, H.; Yamauchi, H.; Uchimaru, Y.; Fujiwara, I.; Haraya, K., **Preparation and gas permeation properties of silicon carbide-based inorganic membranes for hydrogen separation**, *Desalination*, 193 [1-3], 2006; pp. 252–255.

[SYUF2006b]  Suda, H.; Yamauchi, H.; Uchimaru, Y.; Fujiwara, I.; Haraya, K., **Structural evolution during conversion of polycarbosilane precursor into silicon carbide-based microporous membranes**, *J. Ceram. Soc. Japan*, 114 [1330], 2006; pp. 539–544.

[TaBM1992]   Taylor, R.; Brandon, J. R.; Morrell, P., **Microstructure, composition and property relationships of plasma-sprayed thermal barrier coatings**, *Surf. Coat. Technol.*, 50 [2], 1992; pp. 141–149.

[TaWX1999]   Taylor, R. E.; Wang, X.; Xu, X., **Thermophysical properties of thermal barrier coatings**, *Surf. Coat. Technol.*, 120-121, 1999; pp. 89–95.

[TBHB2006]   Torrey, J. D.; Bordia, R. K.; Henager, C. H.; Blum, Y.; Shin, Y.; Samuels, W. D., **Composite polymer derived ceramic system for oxidizing environments**, *J. Mater. Sci.*, 41 [14], 2006; pp. 4617–4622.

[TFHZ2015]   Tang, B.; Feng, Z.; Hu, S.; Zhang, Y., **Preparation and anti-oxidation characteristics of ZrSiO4–SiBCN(O) amorphous coating**, *Appl. Surf. Sci.*, 331, 2015; pp. 490–496.

[Thys2011]   ThyssenKrupp Materials International GmbH, **Werkstoffdatenblatt: Austenitischer hitzebeständiger Stahl 1.4828**, Germany, 2011.

[TLSW2009]   Tan, Y.; Longtin, J. P.; Sampath, S.; Wang, H., **Effect of the Starting Microstructure on the Thermal Properties of As-Sprayed and Thermally Exposed Plasma-Sprayed YSZ Coatings**, *Journal of the American Ceramic Society*, 92 [3], 2009; pp. 710–716.

[ToBo2007]   Torrey, J. D.; Bordia, R. K., **Phase and microstructural evolution in polymer-derived composite systems and coatings**, *J. Mater. Res.*, 22 [07], 2007; pp. 1959–1966.

[ToBo2008a]    Torrey, J. D.; Bordia, R. K., **Mechanical properties of polymer-derived ceramic composite coatings on steel,** *J. Eur. Ceram. Soc.,* 28 [1], 2008; pp. 253–257.

[ToBo2008b]    Torrey, J. D.; Bordia, R. K., **Processing of polymer-derived ceramic composite coatings on steel,** *J. Am. Ceram. Soc.,* 91 [1], 2008; pp. 41–45.

[ToBo2010]    Torrey, J. D.; Bordia, R. K., **Filler systems (bulk components and nano-composites)** in: *Polymer derived ceramics. From nano-structure to applications* (Eds. Colombo, P. et al.). DEStech Publ, Lancaster Pa., 2010.

[Todd2006]    Todd, R. I., **Particulate composites** in: *Ceramic-matrix composites. Microstructure, properties and applications* (Ed. Low, I.-M.). Woodhead, Cambridge, Boca Raton, Fla, 2006.

[TWLG2013]    Tian, H.; Wang, Y. M.; Liu, Y.; Guo, L. X.; Ouyang, J. H.; Zhou, Y.; Jia, D. C., **Dependence of infrared radiation on microstructure of polymer derived ceramic coating on steel,** *Curr. Appl. Phys.,* 13 [1], 2013; pp. 1–6.

[VaGJ2000]    Vaidyanathan, K.; Gell, M.; Jordan, E., **Mechanisms of spallation of electron beam physical vapor deposited thermal barrier coatings with and without platinum aluminide bond coat ridges,** *Surf. Coat. Technol.,* 133-134, 2000; pp. 28–34.

[VCTB2000]    Vassen, R.; Cao, X.; Tietz, F.; Basu, D.; Stöver, D., **Zirconates as new materials for thermal barrier coatings,** *J. Am. Ceram. Soc.,* 83 [8], 2000; pp. 2023–2028.

[Verb1973]    Verbeek, W., **Formkoerper aus homogenen Mischungen von Siliciumcarbid und Siliciumnitrid und Verfahren zu ihrer Herstellung.** Germany, Patent No. DE 2218960 A1, 1973.

[VeWi1974]    Verbeek, W.; Winter, G., **Formkoerper aus Siliciumcarbid und Verfahren zu ihrer Herstellung.** Germany, Patent No. DE 2236078 A1, 1974.

[VGMM2016]    Vaßen, R.; Grünwald, N.; Marcano, D.; Menzler, N. H.; Mücke, R.; Sebold, D.; Sohn, Y. J.; Guillon, O., **Aging of atmospherically plasma sprayed chromium evaporation barriers,** *Surface and Coatings Technology,* 291, 2016; pp. 115–122.

[VJSM2010]    Vaßen, R.; Jarligo, M. O.; Steinke, T.; Mack, D.; Stöver, D., **Overview on advanced thermal barrier coatings,** *Surf. Coat. Technol.,* 205 [4], 2010; pp. 938–942.

[vZPJ2013]    van Ooij, W. J.; Zhu, D. Q.; Prasad, G.; Jayaseelan, S.; Fu, Y.; Teredesai, N., **Silane based chromate replacements for corrosion control, paint adhesion, and rubber bonding,** *Surface Engineering,* 16 [5], 2013; pp. 386–396.

[Wagn1956]    Wagner, C., **Formation of composite scales consisting of oxides of different metals,** *J. Electrochem. Soc.,* 103 [11], 1956; p. 627.

[WaSe2009]    Wang, H.; Sen, M., **Analysis of the 3-omega method for thermal conductivity measurement,** *Int. J. Heat Mass Transfer,* 52 [7-8], 2009; pp. 2102–2109.

[WFPG2003]   Winkler, C.; Flörchinger, P.; Patil, M. D.; Gieshoff, J.; Spurk, P.; Pfeifer, M., **Modeling of SCR deNOx catalyst: Looking at the impact of substrate attributes** in: *SAE technical paper 2003-01-0845*, 2003.

[WGMB2011]   Wang, K.; Günthner, M.; Motz, G.; Bordia, R. K., **High performance environmental barrier coatings: Part II, Active filler loaded SiOC system for superalloys,** *J. Eur. Ceram. Soc.*, 31 [15], 2011; pp. 3011–3020.

[WGMF2013]   Wang, K.; Gunthner, M.; Motz, G.; Flinn, B. D.; Bordia, R. K., **Control of surface energy of silicon oxynitride films,** *Langmuir*, 29 [9], 2013; pp. 2889–2896.

[WHLS2011]   Woiton, M.; Heyder, M.; Laskowsky, A.; Stern, E.; Scheffler, M.; Brabec, C. J., **Self-assembled microstructured polymeric and ceramic surfaces,** *J. Eur. Ceram. Soc.*, 31 [9], 2011; pp. 1803–1810.

[WiVM1974]   Winter, G.; Verbeek, W.; Mansmann, M., **Formkoerper aus homogenen Mischungen von Siliciumcarbid und Siliciumnitrid und Verfahren zu ihrer Herstellung.** Germany, Patent No. DE 2243527 A1, 1974.

[WKSM1990]   Wakai, F.; Kodama, Y.; Sakaguchi, S.; Murayama, N.; Izaki, K.; Niihara, K., **A superplastic covalent crystal composite,** *Nature*, 344 [6265], 1990; pp. 421–423.

[WTQG2012]   Wang, Y. M.; Tian, H.; Quan, D. L.; Guo, L. X.; Ouyang, J. H.; Zhou, Y.; Jia, D. C., **Preparation, characterization and infrared emissivity properties of polymer derived coating formed on 304 steel,** *Surf. Coat. Technol.*, 206 [18], 2012; pp. 3772–3776.

[WUTF2014]   Wang, K.; Unger, J.; Torrey, J. D.; Flinn, B. D.; Bordia, R. K., **Corrosion resistant polymer derived ceramic composite environmental barrier coatings,** *J. Eur. Ceram. Soc.*, 34 [15], 2014; pp. 3597–3606.

[WWPK2004]   Wang, C.; Wang, J.; Park, C. B.; Kim, Y.-W., **Cross-linking behavior of a polysiloxane in preceramic foam processing,** *J. Mater. Sci.*, 39 [15], 2004; pp. 4913–4915.

[WYLF2012]   Wang, Y.; Yang, J.; Liu, J.; Fan, S.; cheng, L., **Fabrication of oxidation protective coatings on C/C–SiC brake materials at room temperature,** *Surf. Coat. Technol.*, 207, 2012; pp. 467–471.

[WyRi1984]   Wynne, K. J.; Rice, R. W., **Ceramics via polymer prolysis,** *Annu. Rev. Mater. Sci.*, 14 [1], 1984; pp. 297–334.

[XSDB2015]   Xu, J.; Sarin, V. K.; Dixit, S.; Basu, S. N., **Stability of interfaces in hybrid EBC/TBC coatings for Si-based ceramics in corrosive environments,** *Int. J. Refract. Met. Hard Mater.*, 49, 2015; pp. 339–349.

[XZZL2014]   Xiao, F.; Zhang, Z.; Zeng, F.; Luo, Y.; Xu, C., **Fabrication of ceramic coatings from polysilazane/aluminum: Effect of aluminum content on chemical composition, microstructure, and mechanical properties,** *Ceram. Int.*, 40 [1], 2014; pp. 745–752.

[YaHO1975]   Yajima, S.; Hayashi, J.; Omori, M., **Continuous silicon carbide fiber of high tensile strength,** *Chem. Lett.* [9], 1975; pp. 931–934.

[YHOO1976]  Yajima, S.; Hayashi, J.; Omori, M.; Okamura, K., **Development of a silicon carbide fibre with high tensile strength,** *Nature,* 261 [5562], 1976; pp. 683–685.

[YiMa2006]  Yimsiri, P.; Mackley, M. R., **Spin and dip coating of light-emitting polymer solutions: Matching experiment with modelling,** *Chem. Eng. Sci.,* 61 [11], 2006; pp. 3496–3505.

[YIYO1981]  Yajima, S.; Iwai, T.; Yamamura, T.; Okamura, K.; Hasegawa, Y., **Synthesis of a polytitanocarbosilane and its conversion into inorganic compounds,** *J. Mater. Sci.,* 16 [5], 1981; pp. 1349–1355.

[YoMI1994]  Yokota, F.; Morikawa, H.; Ishizuka, T., **Determination of impurities in zirconium disilicide and zirconium diboride by inductively coupled plasma atomic emission spectrometry,** *Analyst,* 119 [5], 1994; p. 1023.

[YuKi2013]  Yun, B. K.; Kim, M. Y., **Modeling the selective catalytic reduction of NOx by ammonia over a Vanadia-based catalyst from heavy duty diesel exhaust gases,** *Appl. Therm. Eng.,* 50 [1], 2013; pp. 152–158.

[YYZS2011]  Yang, D.; Yu, Y.; Zhao, X.; Song, Y.; Lopez-Honorato, E.; Xiao, P.; Lai, D., **Fabrication of silicon carbide (SiC) coatings from pyrolysis of polycarbosilane/aluminum,** *J. Inorg. Organomet. Polym.,* 21 [3], 2011; pp. 534–540.

[ZaMK2011]  Zaheer, M.; Motz, G.; Kempe, R., **The generation of palladium silicide nanoalloy particles in a SiCN matrix and their catalytic applications,** *J. Mater. Chem.,* 21 [46], 2011; p. 18825.

[ZFZR2016]  ZFZR - Zentrales Fahrzeugregister, **Jahresbilanz des Fahrzeugbestandes am 1. Januar 2016,** Kraftfahrt Bundesamt, Germany, 2016.

[Zhan2011]  Zhang, D., **Thermal barrier coatings prepared by electron beam physical vapor deposition** in: *Thermal barrier coatings* (Eds. Xu, H.; Guo, H.). Woodhead Pub, Cambridge, 2011; pp. 3–24.

[ZHDY2008]  Zhu, Y.; Huang, Z.; Dong, S.; Yuan, M.; Jiang, D., **Manufacturing 2D carbon-fiber-reinforced SiC matrix composites by slurry infiltration and PIP process,** *Ceram. Int.,* 34 [5], 2008; pp. 1201–1205.

[ZhMi1998]  Zhu, D.; Miller, R. A., **Sintering and creep behavior of plasma-sprayed zirconia- and hafnia-based thermal barrier coatings,** *Surf. Coat. Technol.,* 108-109, 1998; pp. 114–120.

[ZKMM1999] Ziegler, G.; Kleebe, H.-J.; Motz, G.; Müller, H.; Traßl, S.; Weibelzahl, W., **Synthesis, microstructure and properties of SiCN ceramics prepared from tailored polymers,** *Mater. Chem. Phys.,* 61 [1], 1999; pp. 55–63.

[ZSMK2012]  Zaheer, M.; Schmalz, T.; Motz, G.; Kempe, R., **Polymer derived non-oxide ceramics modified with late transition metals,** *Chem. Soc. Rev.,* 41 [15], 2012; pp. 5102–5116.

# 9 ACKNOWLEDGEMENTS

The present work was carried out at the Chair of Ceramic Materials Engineering (CME) of the University of Bayreuth in the Framework of the European Project "FUNEA – Functional Nitrides for Energy Applications", financially supported by the European Commission through the ITN Marie Skłodowska-Curie actions, 7th Framework Programme.

I would like to thank Prof. Dr.-Ing. Walter Krenkel, for giving me the opportunity to develop my thesis at CME and for his support and orientation along the way. To Prof. Dr. rer. nat. Michael Scheffler and to Prof. Dr.-Ing. Uwe Glatzel for kindly accepting the invitation to be part of the examination committee.

I extend my gratitude to Dr. rer. nat. Günter Motz for recognizing my capabilities, for the mentoring and thorough supervision of the work, for the many fruitful discussions, and for the freedom to pursue my own ideas.

Many thanks also to my former colleagues Dr.-Ing. Martin Günthner, Dr. André Prette, Dr.-Ing. Octavio Flores, Dr.-Ing. Martin Seifert and Dr. rer. nat. Thomas Schmalz for sharing their experience and for the discussions and exchange of ideas, which contributed significantly to the present work. Many thanks also to Bernd Martin for his several contributions, especially for his dedication to the planning and construction of the spray equipment for the coating of pipes, to Walter Müller for the assistance with SEM, EDS and XRD investigations, to Sven Scheler for the preparation and characterization of powders and dilatometry measurements, and to all colleagues and HiWis at CME, who somehow contributed to this work.

I also acknowledge the contributions of Prof. Dr. Robert Vaßen and Nikolas Grünwald (thermo-optical dilatometry measurements), of Prof. Dr. Markus Retsch and M.Sc. Bernd Kopera (thermal conductivity measurements by $3\omega$ method), and of Prof. Dr. Rajendra K. Bordia (interpretation of dilatometry results). Many thanks also to Faurecia Emissions Control Technologies, Germany GmbH, especially Prof. Dr. mont. Helmut Wieser, and to Merck KGaA for the cooperation in the Framework of the FUNEA Project.

To my wife Thais, who despite all difficulties, patiently stood by my side along this journey. To my parents for always recognizing the value of a good education and for understanding my choice of moving abroad to pursue my carrier. To my sister for taking such a good care of my parents during my absence. To my dearest friends, who supported and encouraged me.

My most sincere gratitude.

## 10 ANNEXES

### 10.1 List of Symbols, Variables, Chemical Compounds and Abbreviations

*List of symbols and variables*

| | | |
|---|---|---|
| $\varnothing_i$ | mm | Inner diameter |
| $\alpha_c^l$ | [K$^{-1}$] | Linear coefficient of thermal expansion of the coating |
| $\alpha_s^l$ | [K$^{-1}$] | Linear coefficient of thermal expansion of the substrate |
| $\alpha_{af}$ | | Yield after conversion of the active filler |
| $\alpha_{pc}$ | | Ceramic yield after the polymer-to-ceramic transition |
| $\beta_{af}$ | | Density ratio after conversion of the active filler |
| $\beta_{pc}$ | | Density ratio after the polymer-to-ceramic transition |
| $\varepsilon_{af}^l$ | | Linear shrinkage/expansion of the active filler during thermal treatment |
| $\varepsilon_{paf}^l$ | | Linear shrinkage/expansion of the active filler/precursor mixture during thermal treatment |
| $\varepsilon_{pc}^l$ | | Linear shrinkage/expansion of the precursor during the polymer-to-ceramic transition |
| $\kappa$ | [mm$^{-1}$] | Curvature |
| $\lambda$ | [W m$^{-1}$ K$^{-1}$] | Thermal conductivity |
| $\lambda_{a/f}$ | | Air/fuel equivalent ratio or excess air factor |
| $\lambda_c$ | [$\mu$m] | Cutoff length (profilometry) |
| $\rho$ | [g cm$^{-3}$] | Density |
| $\rho_c$ | [g cm$^{-3}$] | Density of the polymer-derived ceramic |
| $\rho_p$ | [g cm$^{-3}$] | Density of the polymer |
| $\sigma_{SD}$ | | Standard deviation |
| $\sigma_{St}$ | [Pa] | Thermal stresses in coatings calculated by Stoney's equation |
| $\sigma_t$ | [Pa] | Thermal stresses in coatings |
| $\nu_s$ | | Poisson's ratio of the substrate |
| $\omega$ | [s$^{-1}$] | Angular frequency of the alternating current |
| $C_P$ | [J g$^{-1}$ K$^{-1}$] | Specific heat capacity |
| $d_c$ | [$\mu$m] | Thickness of coating |
| $d_s$ | [$\mu$m] | Thickness of substrate |
| $D_p$ | [$\mu$m] | Particle diameter |
| $D_t$ | [m$^2$ s$^{-1}$] | Thermal diffusivity |
| $D_{50}$ | [$\mu$m] | Maximum particle size of 50 wt% of the powder |
| $D_{90}$ | [$\mu$m] | Maximum particle size of 90 wt% of the powder |
| $E_c$ | [GPa] | Young's Modulus of the coating |
| $E_s$ | [GPa] | Young's Modulus of the substrate |
| $f$ | [s$^{-1}$] | Frequency |
| $K_0$ | | Modified Bessel function of order zero |

| $l$ | [mm] | Length, indexes 0 and $f$ for initial and final lengths (dilatometry), and $i$ and $e$ for individual and evaluation lengths (profilometry) |
| $m_c$ | [g] | Mass of the ceramic |
| $m_p$ | [g] | Mass of the polymer |
| $M_v$ | | Category of passenger vehicles (European emissions legislation) |
| $M_w$ | [g mol$^{-1}$] | Molecular weight |
| $N_v$ | | Category of cargo vehicles (European emissions legislation) |
| $P$ | [W] | Electric power |
| $q^{-1}$ | [µm] | Penetration depth of a heat wave generated by an alternated current passing through a metallic stripe |
| $r$ | [m] | Radial distance |
| $R_K$ | [mm] | Radius of curvature |
| $R_e$ | [Ω] | Electric resistance |
| $T$ | [K, °C] | Temperature (indexes 0 and $f$ for initial and final temperature) |
| $V_{af}$ | | Volume fraction of active filler |
| $V_{af}^*$ | | Critical volume fraction of active filler |
| $V_{af}^{max}$ | | Maximum packing density of active filler particles |
| $V_v$ | | Volume fraction of voids |

*List of chemical compounds*

| ABSE | Ammonolysed bis(dichloromethylsilyl)ethane |
| AlN | Aluminum nitride |
| $Al_2O_3$ | Aluminum oxide or alumina |
| BCN | Boron carbonitride |
| BN | Boron nitride |
| BSAS | Barium-strontium aluminosilicate |
| CaO | Calcium oxide |
| cBN | Cubic boron nitride |
| $CeO_2$ | Cerium dioxide or ceria |
| $CH_4$ | Methane |
| CO | Carbon monoxide |
| $CO_2$ | Carbon dioxide |
| $Cr_2O_3$ | Chromium oxide or chromia |
| DCP | Dicumyl peroxide |
| hBN | Hexagonal boron nitride |
| $H_2O$ | Water |
| $La_2O_3$ | Lanthanum oxide |
| MgO | Magnesium oxide or magnesia |
| MTES | Methyltriethoxysilane |
| $N_2$ | Nitrogen gas |

| $NH_3$ | Ammonia |
|--------|---------|
| $NO_x$ | Nitrogen oxides |
| PDMS | Polydimethylsiloxane |
| PHMS | Polyhydridomethylsiloxane |
| PHPS | Perhydropolysilazane |
| PMMS | Polymethoxymethylsiloxane |
| PS | Polystyrene |
| R | Side groups of preceramic polymers |
| $SO_2$ | Sulfur dioxide |
| $Si_3N_4$ | Silicon nitride |
| SiAlCN | Silicoaluminum carbonitride |
| SiAlCO | Silicoaluminum oxycarbide |
| SiBCN | Silicoboron carbonitride |
| SiBCO | Silicoboron oxycarbide |
| SiC | Silicon carbide |
| SiCN | Silicon carbonitride |
| SiCNO | Silicon carboxynitride |
| SiCO | Silicon oxycarbide |
| $SiO_2$ | Silicon dioxide or silica |
| $SiO_x$ | Silicon oxides |
| $TiSi_2$ | Titanium disilicide |
| $Y_2O_3$ | Yttrium oxide or yttria |
| YSZ | Yttria-stabilized zirconia |
| 3YSZ | Zirconia stabilized with 3 mol% of yttria |
| 4YSZ | Zirconia stabilized with 4 mol% of yttria |
| $ZrO_2$ | Zirconium dioxide or zirconia |
| $ZrSi_2$ | Zirconium disilicide |

*List of Abbreviations*

| AC | Alternated current |
|----|--------------------|
| AFCOP | Active-filler-controlled pyrolysis |
| APS | Air plasma spraying |
| CI | Compressive ignition (engine) |
| CMC | Ceramic matrix composite |
| CTE | Coefficient of thermal expansion |
| CVD | Chemical vapor deposition |
| EBC | Environmental barrier coating |
| EB-PVD | Electron-beam physical vapor deposition |
| EDS | Energy-dispersive X-ray spectroscopy |
| EEA | European environment agency |
| EGR | Exhaust gas re-circulation |

| FSZ | Fully stabilized zirconia |
| GDOES | Glow discharge optical emission spectroscopy |
| IRRAS | Infra-red reflection absorption spectroscopy |
| HC | Hydrocarbons |
| IC | Internal combustion |
| LPG | Liquefied petroleum gas |
| MEMS | Micro-electromechanical systems |
| NEMS | Nano-electromechanical systems |
| NSC | $NO_x$ storage catalyst |
| PDC | Polymer-derived ceramic |
| PI | Positive ignition (engine) |
| PIP | Precursor infiltration and pyrolysis |
| PM | Particulate matter |
| PSZ | Partially stabilized zirconia |
| PVD | Physical vapor deposition |
| RTD | Resistance-temperature detector |
| SCR | Selective catalytic reduction |
| SEM | Scanning electron microscopy |
| SPS | Spark plasma sintering |
| TBC | Thermal barrier coating |
| TGA | Thermogravimetric analysis |
| TGO | Thermally grown oxide |
| TSZ | Tetragonal-stabilized zirconia |
| VOC | Volatile organic compounds |
| XPS | X-ray photoelectron spectroscopy |
| XRD | X-ray diffraction |

## 10.2 Publications

Barroso, G.; Li, Q.; Motz, G.; Bordia, R. K., **Polymer-derived ceramic and ceramic-like coatings: Innovative solutions for real problems,** *Am. Ceram. Soc. Bull.,* 96 [3], 2017; pp. 42–49.

Tangermann-Gerk, K.; Barroso, G.; Weisenseel, B.; Greil, P.; Fey, T.; Schmidt, M.; Motz, G., **Laser pyrolysis of an organosilazane-based glass/ZrO2 composite coating system,** *Mater. Des.,* 109, 2016; pp. 644–651.

Parchoviansky, M.; Barroso, G.; Petrikova, I.; Motz, G.; Galuskova, D.; Galusek, D.: **Polymer derived glass ceramic layers for corrosion protection of metals,** In (Lin, H.-T. et al. Eds.) Advanced and refractory ceramics for energy conservation and efficiency: A collection of papers presented at CMCEE-11, June 14-19, 2015, Vancouver, BC, Canada. Wiley, Hoboken, New Jersey, 2016.

Barroso, G.; Kraus, T.; Degenhardt, U.; Scheffler, M.; Motz, G., **Functional coatings based on preceramic polymers,** *Adv. Eng. Mater.,* 18 [5], 2016; pp. 746–753.

Coan, T.; Barroso, G. S.; Machado, R.; Souza, F. S. de; Spinelli, A.; Motz, G., **A novel organic-inorganic PMMA/polysilazane hybrid polymer for corrosion protection,** *Prog. Org. Coat.,* 89, 2015; pp. 220–230.

Barroso, G. S.; Krenkel, W.; Motz, G., **Low thermal conductivity coating system for application up to 1000°C by simple PDC processing with active and passive fillers,** *J. Eur. Ceram. Soc.,* 35 [12], 2015; pp. 3339–3348.

Coan, T.; Barroso, G. S.; Motz, G.; Bolzán, A.; Machado, R. A. F., **Preparation of PMMA/hBN composite coatings for metal surface protection,** *Mat. Res.,* 16 [6], 2013; pp. 1366–1372.

## 10.3 Curriculum Vitae

### Personal Information

Name:            Gilvan Barroso

Address:         Leuschnerstr. 56, 95447 Bayreuth, Germany

Date of birth:   06.07.1988

Place of birth:  Blumenau, Brazil

Marital status:  Married

### Education

| | |
|---|---|
| 2005 | Escola Técnica do Vale do Itajaí – ETEVI |
| | Blumenau, Brazil |
| Mai 2006 – Aug. 2011 | Undergraduation in Chemical Engineering at the Federal University of Santa Catarina |
| | Florianópolis, Brazil |

### Professional Experience

| | |
|---|---|
| since Dec. 2011 | Research assistant at the Chair of Ceramic Materials Engineering (CME) of the University of Bayreuth |
| | Bayreuth, Germany |
| | Topic of research: Development of coatings and composites based on preceramic polymers |

www.ingramcontent.com/pod-product-compliance
Lightning Source LLC
Chambersburg PA
CBHW060450240326
41598CB00088B/4334